KILLERS
OF THE WILD

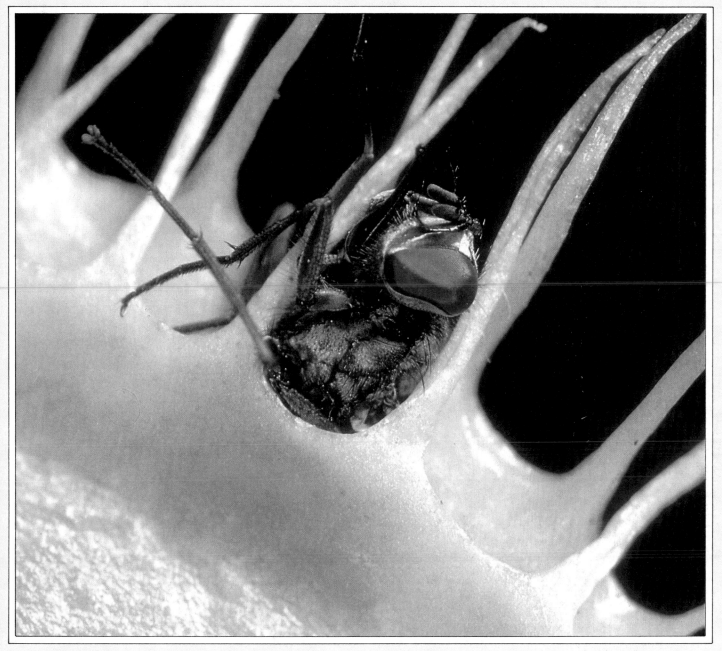

Firmly held in the 'jaws' of the Venus' fly-trap, a fly will be digested to feed the plant.

Fastest animals on the African plains, some cheetahs have killed a zebra for a good feed.

KILLERS
OF THE WILD

Author: Michael Chinery
Consultants: Dr Pat Morris/Dr Diane Hughes

a Salamander book

Published by Salamander Books Limited
LONDON

A Salamander Book

Published by Salamander Books Ltd.,
Salamander House,
27 Old Gloucester Street,
London WC1N 3AF,
England.

Distributed in the United Kingdom by
New English Library Ltd.

Distributed in Australia/New Zealand
by Summit Books, a division of Paul
Hamlyn Pty Ltd., Sydney, Australia.

A red fox pounces on
a mouse in the grass.

The Galapagos hawks clearly show the hooked, flesh-tearing beak typical of birds of prey.

Credits

Editor: Geoff Rogers
Designers: Roger Hyde   Rod Teasdale

Colour reproductions:
         Culver Graphics Litho Ltd.
         Process Colour Centre Ltd.
         Tenreck Ltd., England.

Filmset: SX Composing Ltd., England.

Printed in Belgium by
Henri Proost et Cie, Turnhout.

A giant anteater breaks into a termite mound and licks up its prey.

The Author

Michael Chinery's first childhood outing was, so he has been told, a visit to the zoo at the tender age of one. It must have made an impression, for he has been looking at animals and plants ever since. As a schoolboy, he always had a room full of caterpillars, lizards, fish, and various other small animals, and spent a good deal of the school holidays chasing butterflies or paddling in the local ponds and streams in search of water scorpions, hydra, and other strange creatures.

After graduating from Cambridge in 1960 with a degree in natural sciences and a diploma in anthropology, Michael Chinery devoted himself to writing about wildlife. He is now widely known as a writer of natural history books, with a guide to the insects of Europe among his many published works. He is also a regular broadcaster on the BBC's *Wildlife*

Beautifully camouflaged in a flower, a crab spider has killed an unsuspecting honeybee.

programme and a lecturer for the Field Studies Council, taking groups of people out into the countryside and helping them to identify and learn more about the animals and plants around them. A lot of his time is spent watching and photographing the activities of the many small animals in his garden – always a good hunting ground for the naturalist – and in the surrounding fields and hedgerows of the Suffolk countryside.

Further afield, Michael Chinery has watched and photographed animals, especially insects, in many parts of Europe, in West Africa, and in the Far East. He has incorporated many of his first-hand observations into this book, and a number of his photographs are also scattered through the book, especially in the invertebrate sections.

The Consultants

Dr. Pat Morris is a lecturer in zoology at London University and has travelled widely in pursuit of his natural history interests. He is a frequent contributor to nature radio programmes.

Dr. Diane Hughes studied zoology at London University, then researched disease in wildlife. She has visited many game parks in Europe, Africa and India to watch animals in their natural habitat.

A tawny owl, its talons spread, swoops silently onto its prey.

Contents

Left: Wild hunting dogs bring down a wildebeeste.

Right: Cyanea — a beautiful but deadly jellyfish.

A young pike has broken cover to catch a stickleback in its already powerful, toothy jaws.

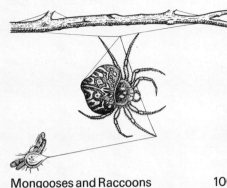

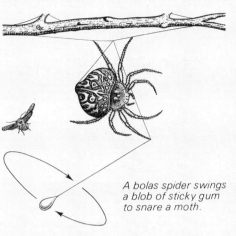

A bolas spider swings a blob of sticky gum to snare a moth.

Above: A swift scoops up midges in its beak.

Right: Australian Aborigines trap a baby shark.

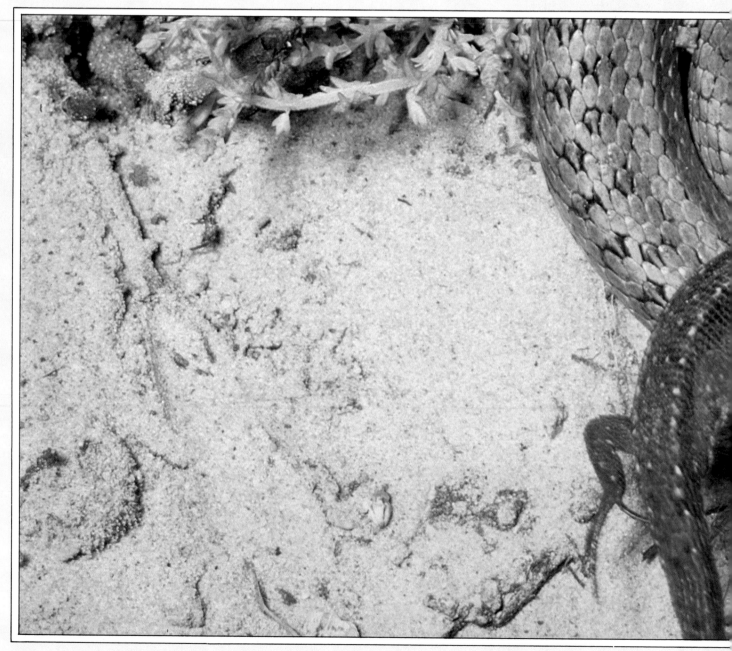

A European smooth snake begins to swallow a common lizard that it has already strangled.

Foreword

Feeding is one of the basic features of all living things and, over the millions of years that animals have been living on the earth, they have learned to exploit every available source of food. This has meant that many have become predators – killers of other animals. For many people, this will conjure up visions of bloody battles between lions and zebras or antelopes, but such spectacular killings are only a very small part of the whole predatory scene. Predation is not confined to large animals, and many examples can be found simply by walking round an ordinary garden, or even by looking around the house: there can be few houses without spiders, and spiders are among the most efficient predators of insects and other small creatures. Other familiar predatory activities that can be seen around the garden include the song

thrush hammering snails on a stone, the blackbird tugging a worm from the lawn, and the swallows and bats swooping through the air to snatch insects in full flight. The aphid-munching ladybird is another well-known predator. And then, of course, there is the domestic cat that regularly brings home mice and voles, in which case it is usually congratulated, or birds, in which case it tends to be scolded.

These familiar killers are no less lethal to their victims, or prey, than the lion is to the antelope, and it is only their lack of size that prevents them from commanding the same sense of awe that surrounds the lion and the other big predators. But anyone who cares to examine some of these smaller killers in detail cannot fail to admire their techniques. There is the spitting spider, which actually fires sticky threads from its jaws to pin down its prey;

the archer fish, which fires a hail of water 'bullets' at insects above the surface; the sea anemone that fires 'harpoons' into its prey; and the glow-worm that lures prey to its death by means of flashing lights. And then there are the bats that detect and track down their insect prey with radar, and the electric fishes – not particularly small – that navigate and find food by surrounding themselves with electric fields.

By comparison with some of these elaborate techniques, the stalking and charging of the lion, the leopard, and the wolf seem rather crude, but the predator's methods must always be judged on results, and if they fill the animal's belly they must be considered to be successful.

A pack of hunting dogs or other large carnivores chasing and pulling down their prey are often described as being blood-

A hunting wasp drags a paralyzed bush cricket to its burrow as food for its young.

thirsty creatures, but this is a very anthropomorphic view and the description is not justified. The animals are not really any more eager to shed blood than the blackbird tugging its worm from the soil: they are all simply obeying an instinct to catch their normal food, and they rarely kill more than they need. Only in exceptional circumstances, such as captivity or overcrowding, do the predators indulge in excessive killing.

If predators killed unnecessarily in the wild, they would soon run out of food and die. Predation is an integral part of nature's complex web, with numerous controls and feed-back systems ensuring that wildlife remains in balance with its habitat. An increase in predator population leads to a decrease in food supplies, and prey becomes more difficult to find. Many predators die or else leave the area, and the prey population can then return to normal, followed later by the predator population. Such fluctuations are most common in the poorer habitats, such as the tundra, where there are relatively few animal species. The population of the snowy owl, for example, undergoes marked fluctuations as a result of variations in the population of the lemming, which is almost its only prey. In more complex habitats, where the food chains have more cross links, fluctuations in the numbers of one prey species do not have such marked effects because the predators can turn to other prey. Predator and prey are thus kept in a nice balance, and predation can actually benefit the prey species by weeding out the weakest individuals in each generation and thus improving the vigour of the remaining population – a far cry from the 'blood-thirsty' image often associated with predation.

Only a small selection of predatory animals are described in this book, but I have tried to include all the major groups of predators and all the major predatory techniques, from the simple, but highly effective ambush of the crab spider, through the elaborate and very efficient traps of the spiders, to the high-speed methods of the cheetah, the peregrine, and the barracuda. Next time the cat brings in a mouse or a bird, you may still feel pleased or angry, but I hope you will be a little more aware of just what has been going on in your garden.

Michael Chinery

The Predatory Life

Killing to eat is a perfectly normal function for many animals, and numerous strategies have been evolved to enable the predators to capture and kill their prey. The kill may sometimes appear cruel and blood-thirsty to human eyes, but it is really very efficient, and little is wasted: when these hyenas have had their fill the waiting vultures will move in to clear up the scraps.

The Predatory Life

Every animal on the earth needs a supply of food if it is to survive. The food provides it with body-building materials for growth, and with the energy it needs to move about and fulfil its daily routines — including the collection of more food. The size of an animal clearly affects the amount of food that it requires. A lion, for example, needs a lot more meat to keep it going than a small domestic cat. But size is not the only factor involved: activity is important, because activity uses energy and therefore requires more food. A parasitic worm lying in the body of another animal needs far less food than a worm of similar size that swims freely in the water. Similarly, a cold-blooded frog needs far fewer insects to keep it going than a bird of similar size. The bird needs food and energy simply to keep its body at the right temperature and it must eat regularly, but the frog does not have to maintain a high body temperature and it can go without food for long periods.

The sources of food

Unlike plants, which can make their own food from water, carbon dioxide, and

The cheetah — a true carnivore
Above The cheetah is a member of the Carnivora and displays many of the basic features of hunters — good eyes and other senses, great speed, and powerful weapons in the form of cutting and stabbing teeth.

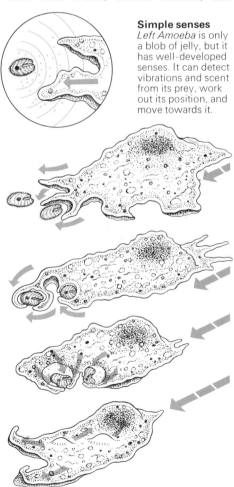

Simple senses
Left Amoeba is only a blob of jelly, but it has well-developed senses. It can detect vibrations and scent from its prey, work out its position, and move towards it.

Amoeba – a microscopic predator
Above Detecting prey in the surroundings — in this instance two minute paramecia — the amoeba flows towards them (top) and com-

pletely surrounds them (centre diagrams). Enclosed within the amoeba body, the prey is carried round and quickly digested.

simple minerals, the animals have to obtain ready-made food consisting of proteins, carbohydrates, and other complex organic substances. They get it basically by eating plants or by eating other animals, and on this basis we can recognize four major groups of animals: herbivores, carnivores, omnivores, and scavengers. The herbivores are the vegetarians — the elephants and antelopes, the rabbits, the voles, the hordes of plant-eating insects, and many others. The carnivores are the flesh-eaters — animals that catch and eat living prey. We usually think of the big cats, the wolves, and the weasels as typical carnivores, because these all belong to the order of mammals known as the Carnivora, but in its widest sense the word carnivore embraces an immense range of predatory animals, from the great killer whale down to microscopic protozoans that engulf their even smaller brethren. Eagles, crocodiles, dragonflies, and spiders are all carnivores, but to avoid confusion with the Carnivora it is best to describe them all as predators — animals that prey on other animal species, large or small.

Omnivores are animals that feed regularly on both plant and animal food. We ourselves are typical omnivores. Other examples from the world around us include the brown and black bears, the baboons, and the European blackbird (*Turdus merula*). The scavengers are those animals that obtain the bulk of their food from dead organisms, both plant and animal. The best known are

the vultures that gather round large corpses on the African plains and elsewhere, but a great many other animals work, full-time or part-time, for Nature's public cleansing department. These include the jackals, several gulls, and huge armies of beetles and cockroaches that remove leaf litter, dung, and even bones and skin left behind by the predators.

In this book we are clearly interested primarily in the predators — those animals that habitually kill for food — but, as we shall see, we cannot completely ignore the importance of either the omnivores or the scavengers.

Food and energy chains

Whatever food source an animal utilizes for its energy supplies, we can trace that energy back to a plant in a very few steps. Take, for example, a sparrowhawk catching and eating warblers in a wood: the warblers that fuel the sparrowhawk's engines get their energy from caterpillars and other small insects, and these insects feed on plants. This is an example of a simple food chain, or energy chain.

Many chains are even shorter — grass→rabbit→fox is a typical example — and some are longer. Leaf→caterpillar——→frog——→snake——→snake eagle is not an uncommon chain in tropical

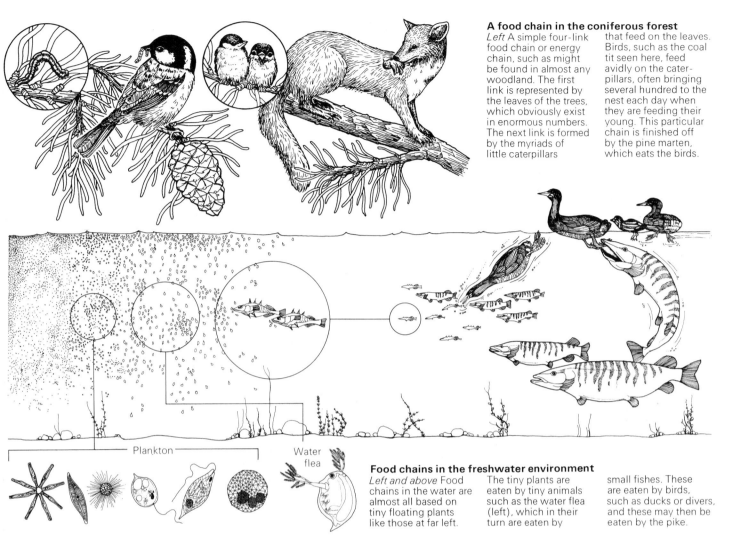

A food chain in the coniferous forest
Left A simple four-link food chain or energy chain, such as might be found in almost any woodland. The first link is represented by the leaves of the trees, which obviously exist in enormous numbers. The next link is formed by the myriads of little caterpillars that feed on the leaves. Birds, such as the coal tit seen here, feed avidly on the caterpillars, often bringing several hundred to the nest each day when they are feeding their young. This particular chain is finished off by the pine marten, which eats the birds.

Food chains in the freshwater environment
Left and above Food chains in the water are almost all based on tiny floating plants like those at far left. The tiny plants are eaten by tiny animals such as the water flea (left), which in their turn are eaten by small fishes. These are eaten by birds, such as ducks or divers, and these may then be eaten by the pike.

regions, and there might be another link, such as a dragonfly, between the caterpillar and the frog. Another link could be added if we count the parasites on or in the eagle, but normal food chains rarely have more than five links. This is because energy is lost at each stage. An adult vole, for example, might contribute just 40gm (1·4oz) to a kestrel's dinner, but before it can do this it must eat at least 10kg (22lb) of grass and other vegetable food. One 40gm (1·4oz) vole will satisfy a kestrel for one day, and this means that at least 3,650kg (8030lb) of plant material is needed to sustain just one kestrel for a year, although the kestrel itself may weigh no more than 200gm (7oz)!

Any additional links in the chain would require enormous increases in productivity lower down the line. Imagine a super-predator weighing about 1kg (2·2lb) and eating just one kestrel each day. If the average age of the captured kestrels were six months, the number of voles required to sustain one super-predator would be over 66,000 per year, and the weight of grass needed to nourish these would be of the order of 660,000kg (1,452,000lb) – all to support an animal weighing just one kilogram. It is not surprising that few food

chains have more than five links.

All food chains must start with green plants, because only these can make organic food from inorganic materials and incorporate energy into it. The food making process is called photosynthesis and, if we exclude a few simple bacteria, it is vital to all life on the earth. During photosynthesis the plants trap the energy of sunlight with the aid of the green pigment called chlorophyll. The energy is used to convert water and carbon dioxide into glucose sugar, which is one of the basic molecules of life. It contains energy, and it can be combined with minerals to form all the essential components of living matter.

When a plant is eaten by a herbivore, the glucose and other materials, together with their bound-in energy, pass to the animal. The energy is released for various purposes when the animal 'burns' the glucose in its body during the process known as respiration. This is chemically the reverse of photosynthesis, for it releases water and carbon dioxide again as well as energy, although the energy is not usually in the form of light. Because energy is always being used in the body, the amount contained in a body at any one time – when it is eaten, for example – is only a fraction of the

energy that has been taken in during its life. This explains the enormous loss of energy with each link of a food chain.

The pyramid of numbers
As a result of their unique ability to make food, the plants are often referred to as the primary producers. The herbivorous animals that eat them are then called the primary consumers and the predators that form the next link in the chain are called secondary consumers. There may also be tertiary and even quaternary consumers, as we have seen in the chain ending with the snake eagle. The primary producers – the plants – clearly must outweigh the primary consumers that feed on them. Likewise, the primary consumers must outweigh the secondary consumers, and so on. A food chain can therefore be represented as a pyramid – the so-called pyramid of numbers – with the base representing the plants.

Primary producers may actually outnumber the animals that eat them, as in the grass→rabbit→fox food chain, but they are not always literally more numerous than the primary consumers. Trees, for example, are far less numerous than the caterpillars that nibble their leaves, and here we must take into account the biomass – the total weight of the living material. Although there are more insects than trees, the biomass of the trees in the habitat far exceeds that of the insects feeding on them. Higher up the pyramid we really can deal in numbers, for there are obviously more voles in a meadow than there are kestrels waiting to pounce on them. There are also far more caterpillars in the trees than there are warblers searching for them.

The animals at the top of the pyramid have no real enemies and they are called top predators, but the energy flow does not stop there. When these animals die their bodies are broken down by hordes of scavengers and decomposers, and the energy that was locked up in their tissues is gradually dissipated as it passes along new food chains.

If we consider just the predators in a food chain or pyramid of numbers, we find that, in general, the higher we go the larger the animals become. This seems obvious, because predators usually are larger than their prey, but there are some exceptions. Wolves and hunting dogs, for example, often kill animals considerably larger than themselves, but these predators go in for group hunting and we should really compare the weight of the pack with the weight of the prey.

Most predators concentrate on prey within certain size limits, and may even be frightened of creatures that exceed these limits. Many praying mantises, for example, go into a defensive attitude when faced with extra large moths. Creatures well below the limits are generally ignored. A lion, for example, takes little or no notice of a mouse, apparently realizing that the reward would not be worth the effort of catching it. There are, however, a few examples of predators taking disproportionately small prey. The anteater and the aardvark are two such examples, but it must be remembered that they attack whole colonies of ants and termites and that one assault can reap a large total reward.

The whalebone whales also take tiny prey — crustaceans, called krill, that float in the surface layers of the sea. The whales are actually cutting out the middle links of a normal food chain. There is much to be gained in short-cutting the chains in this way if large quantities of small prey can be obtained in one place because energy losses are reduced as well. Far more whales can exist on a given volume of krill by direct predation than would be possible if the krill first had to pass through fishes.

The web of life
Dozens or even hundreds of different food chains can exist in one small habitat, but if we look at them carefully

The pyramid of numbers
Below The food chains on the European tundra begin with the many lichens, grasses, and other low-growing plants. These are the primary producers. The primary consumers are the rodents and hares and other herbivores. Much less numerous are the secondary consumers, such as the Arctic fox and stoat. Top predator, less common again, is the wolf.

we can see that they are all linked together to form a very complex system that we can call the web of life. Herbivorous animals often have very restricted diets — the koala, for example, eats nothing but the leaves of eucalyptus trees, and many insects confine themselves to one particular food plant — but predatory animals are not usually so fussy. Many certainly do have preferences, but most will accept a wide variety of prey animals and they will eat virtually anything when they are really hungry. One of the most striking examples of this was a crayfish which ate a hibernating snake.

Starting with our simple grass⟶ rabbit⟶fox food chain, we can see how the web of life builds up through the numerous predator/prey relationships. The fox is the top predator in a great many food chains, of which we can cite just three here: grass⟶vole⟶fox; lettuce⟶snail⟶thrush⟶fox; decay-

Owl

Shrew

Vole

Fox

Buzzard

Ground beetle

Rabbit

Badger

Blackbird

Slug

Death and decay

Fruit

Grass and other vegetation

Earthworm

Decay

A complex food web
Above The simple grass-rabbit-fox food chain can be linked with many other food chains in the habitat by bringing in the many other animals that the fox eats and the many other predators that catch rabbits. Just a few of these links are shown in the diagram, and they do not include any of the other plants that the rabbit eats, but already the diagram, which represents part of a grassland food web, is very complex. The more complex the food web of a habitat, the more stable the community: if one link should disappear, the consumers can turn to other sources of food.

Predator and prey
Left A bobcat chases a snowshoe hare at high speed. The bobcat is a very versatile predator, with many types of prey.

ing leaves→earthworm→hedgehog→ fox. In these three examples we have linked five other animals to the rabbit by way of the fox. We can join another lot in by considering some of the rabbit's other predators, such as the buzzard, the badger, the stoat, and various owls. If we then construct food chains around these predators, showing all their possible prey and the prey's food as well, we will have hundreds of plants and animals all linked together in the web.

Recipe for success
It is obvious from the foregoing that the fox can thrive on many different kinds of prey. If it had been completely dependent on rabbits, it would have disappeared from Europe when myxomatosis killed nearly all the rabbits in the 1950s. But it suffered very little because it was able to turn to voles and other prey. The fox is clearly a very successful predator. It is sufficiently specialized to be able to track down and kill its prey efficiently, and yet it remains sufficiently flexible in behaviour to be able to deal with several kinds of food. The Everglades kite is another very efficient predator, but it is much more specialized than the fox and utterly dependent on one source of food – a particular kind of water snail. Efficient it may be, but it cannot really be called successful because, with only one food chain to support it, it is confined to one type of habitat and is very vulnerable to environmental change.

Red in tooth and claw
Tennyson's reference to Nature being red in tooth and claw perhaps gives the impression that predators are cruel and blood-thirsty creatures, eager to kill at every opportunity. Some kills, such as those of the hunting dogs, certainly are bloody affairs, but the dogs are merely following their instincts to feed and survive and we cannot condemn them for this. The hunting dog kill is dramatic, but we cannot really say that it is cruel, any more than we can say that a frog is cruel when it gulps down a fly or a caterpillar is cruel when it nibbles a leaf.

Killing is a necessary part of the complex balance of nature, exerting a considerable effect on the stability of the populations of herbivorous species. There are many examples of population explosions among the herbivores following the removal or reduction of the predators. Hippos increased enormously in parts of Africa when most of the crocodiles were shot, because the crocodiles used to eat young hippos. White-tailed deer increased in many parts of North America as the wolf population was reduced, but one of the most striking examples concerns the mule deer of the Kaibab Plateau in Arizona.

At the beginning of the 20th century the area contained perhaps 4,000 deer and a wealth of predators, such as wolves, pumas, and bob-cats, which held them in check. In 1906 the area became a game reserve and the predators were killed in their hundreds – 600 pumas destroyed in ten years, 3,000 coyotes wiped out in 16 years, and all the wolves gone within 20 years. The deer responded rapidly. There were 100,000 of them in 1924, and they were rapidly destroying their own habitat. Thousands died in the next few years through lack of food, and by 1931 the population had fallen to 20,000. But this was still too many for the good of the forest, and man had to take over the job of population control with the gun – a job that the predators had previously done very effectively and at no cost at all.

As well as controlling populations, the predators may actually improve the health of the prey populations by weeding out the less-fit individuals and leaving the fitter and healthier ones to breed. A cheetah sprinting after antelopes on the plains is clearly going to have a better chance of catching a sickly animal than a fit one. Hunting dogs also have more success against weaker prey; the weaker individuals lag behind the herd and are relatively easy targets for the dogs. This is natural selection at work.

Observations in America have shown that muskrats strong enough to hold and defend territories are rarely killed by mink or foxes. These predators concentrate on the homeless muskrats, which are always the weakest ones. There is an interesting parallel here with some homing pigeons killed by peregrine falcons in Germany. Fifteen out of seventeen pigeons killed were in unfamiliar territory, and therefore less confident and probably flying more slowly than usual. It has also been shown that one third of the roach caught by cormorants in Holland are infected with tapeworms, although only about one sixteenth of the total roach population is so affected. Clearly the cormorants have more success in catching the infected and sickly fishes, but in none of these examples should it be thought that the predators initially select the weaker prey for attack. The weaknesses show up only when the chase begins, and nature then takes its course. Only the fittest survive.

A fox that gets into a chicken run will often kill all the chickens – far more than it can possibly eat. A weasel and a cat may do the same if they are surrounded by small prey in an enclosed space, and this has given them their unjustified reputation of blood-thirsty killers. Such behaviour is abnormal, and rarely, if ever, happens in the wild. The weasel's attack, for example, is triggered off by movement, and when it has caught one prey animal in the wild the other potential victims have gone to ground. Only in an enclosed space will the weasel carry on killing for no apparent reason.

Paul Leyhausen has investigated this behaviour with cats and has come up with an interesting explanation. He gave a cat a succession of live mice, and found that the first few mice were killed and eaten. The cat was then full, but it continued to catch and kill the mice that were provided, although it did not eat them. As further mice were provided, the cat gave up killing them, although it still continued to stalk and catch them. In the end, the cat became tired and did no more than stalk the mice.

Leyhausen's explanation is that the different facets of prey-catching behaviour – stalking, catching, killing, and eating – require different degrees of motivation. The cat requires very little motivation to begin stalking, but a much higher level of motivation – the pangs of hunger, perhaps – before it will eat. But why should a satiated cat even bother to stalk mice? Consideration of the cat in the wild provides the answer. The cat is not surrounded by easy prey in the wild, and it must always be on the lookout for a meal and ever-ready to begin stalking. Only a small percentage of stalks end successfully, and if a cat began stalking only when it was hungry it might starve before it had any success. But if it is always ready to stalk it has a chance of catching food at any time. The higher levels of motivation needed for the later stages of the chase mean that the cat will carry on to the finish only when it is really hungry. It can abort its mission at any time, but under natural conditions it is unlikely that a cat will be faced with so much prey that it is stimulated to kill without eating.

Predatory methods

Every creature that catches another for food is technically a predator, and each has its particular method of capturing its prey. Some of these methods are very simple – the blue tit pecking insects from a tree trunk is a good example – but they still involve efficient senses to detect the prey and the right sort of equipment to catch it. At the other end of the scale, there are some really amazing methods, involving radar, electricity, and trip-wires to detect, trap, and kill the prey.

We can divide the predators into three main groups; the ambushers, the trappers, and the hunters. The ambushers lie in wait for their prey and depend very much on camouflage to conceal them from their potential victims. Well-known examples include the mantises and the crab spiders. Although the ambushers instinctively position themselves in places that are likely to be visited by their prey, there is an element of chance in this method, and most of the ambushers are able to go in search of food if none arrives in a certain time.

Relatively few animals actually make traps for their prey. The only well-known ones are the spiders, which use silk to make webs of various kinds. Quite a number of animals use lures, however, to attract prey within range of their jaws. The angler fishes and some of the fireflies are good examples of this technique.

The majority of predators actually hunt their prey. They track it down with eyes, ears, nose, and various other senses – some of them very surprising – and then give chase. The chase may be just that – a fast run, swim, or flight in an attempt to overhaul the prey by sheer speed – or it may be a stealthy affair, with the predator creeping slowly up to

The hidden predator
Left The crab spider is a master of the ambush, usually lurking on the ground or on vegetation to grasp unwary insects in its stout front legs and to deliver a paralyzing bite with the efficient poison fangs.

The golden eagle
Right The great hooked beak of the golden eagle, as in other birds of prey, is used to tear the flesh from the prey – in this example, a jackrabbit. The talons are extremely powerful and used to kill the prey.

Bumping into food
Below The ladybird is a slow, but nevertheless effective predator. It does not need speed to catch the aphids on which it feeds – just an ability to recognize them when it bumps into them and jaws with which to chew them thoroughly.

The delicate trap
Above The web of the spider *Frontinella pyramitela* hangs from many scaffold threads, which impede small insects and make them fall into the hammock.

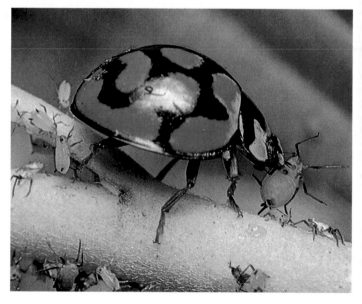

take its prey by surprise. Exponents of the fast chase include the cheetah, the peregrine falcon, and the barracuda. The stealthy approach is confined largely to land-dwelling animals and is beautifully shown by most of the cats, by the chameleons, and also by some of the jumping spiders.

Trout sometimes leap out of the water to snatch mayflies and other insects flying over the surface, and the caracal sometimes leaps up to snatch birds from the air, but in general aquatic creatures hunt only in the water and land-based predators confine themselves to prey on the ground. The birds and the bats, however, have the best of all worlds: there are species that capture prey in mid-air, some that snatch prey from the ground at the end of a high-speed dive

or glide, and some that scoop food from the water as they fly over it or dive into it.

The predator's armoury

The first requirement for a predator is some form of sensory equipment to enable it to find or recognize its prey. In the typical hunters, such as the birds of prey, the sharks, and the cats, this equipment works at a distance to pick up movements, scents, or sounds, but there are also predators that literally have to bump into their prey before they recognize it with organs of taste and touch. The ladybirds and the centipedes are hunters of this kind.

Having detected potential prey, and caught it by one of the methods already described, the hunter must overpower and kill the victim. The expression 'red

in tooth and claw' certainly applies here, for among the vertebrate predators it is almost always the teeth or the claws that do the damage, with the beak taking the place of the teeth among the birds. No special weapons are used when the prey is relatively small in comparison with the predator; the prey is simply grabbed in the jaws and swallowed – as when a lizard snaps up a slug or a bird grabs a caterpillar. But larger prey has to be subdued or completely killed before it can be eaten. The constricting snakes have no weapons in the accepted sense, but their strength is a formidable weapon and they kill their prey very effectively by coiling themselves around it and suffocating it. Most other vertebrates predators use teeth, beaks, and claws to kill their prey and tear it to pieces. Birds of prey kill with their powerful and sharp talons, which sink deep into the flesh of their victims, and they then use their beaks to tear the flesh. Sharks and mammals use their teeth for killing and tearing, while water birds generally kill with their beaks and swallow their food whole. Much prey is swallowed alive.

Poison is used as a weapon by a large number of predators, the best known being some of the snakes. Their poison is produced in special salivary glands and it enters the wounds made by the teeth. The spiders and many other invertebrates also use poison to subdue their prey, but surely the most unusual weapon in the predators' armoury is the electricity used by various fishes, both to detect and to kill their prey.

Instinct and learning

The invertebrate animals are capable of a certain amount of learning, but the vast majority of their actions, including their predatory behaviour, are controlled by instincts – inborn behaviour patterns handed down from generation to generation. The spider is not taught how to make its web, and yet it generally makes an excellent job of it just by following its instincts. Those spiders that 'lose' their instructions fail to make a satisfactory web as a rule, fail to catch food, and therefore die and fail to give rise to another generation of inefficient spiders. But failure to receive the correct instructions is not always bad; modified instructions sometimes lead to improvements and they are passed on to succeeding generations. This is, of course, the basis of evolution.

Vertebrate animals also inherit instinctive behaviour patterns, but they have a much greater capacity for learning than the invertebrates and they can very often modify their behaviour according to the circumstances. This is especially true of mammals, in which the young

always spend their early days with their mother and often with their father as well. Predatory mammals have a longer upbringing than most others and the youngsters play with each other a good deal. They may also play with their parents, and they are usually taught how to hunt. Most of them are born with the instinct to chase after something, but they have to learn the finer points of catching and killing from their parents. This can lead to variations in habits and also in foods between different populations. If one predator discovers a new trick or technique that is useful in hunting, it will pass it on to its offspring during the 'teaching' period and the new habit will gradually spread through the population.

Birds also learn quite well, although there is not quite so much opportunity for parents to pass knowledge to their offspring. One marvellous example of avian learning comes from a park in Florida, where a heron had noticed that fishes came to the surface when people threw food pellets into the water. It was not long before the heron, too, was throwing pellets into the water, and catching considerably more fish as a result!

The ability of birds and mammals, and other vertebrates, to learn and to modify their behaviour accordingly means that

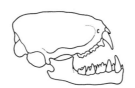

Teeth to crunch
Above The hoary bat (*Lasiurus cinereus*) has many small, sharp teeth with which to crunch up horny insect skeletons.

Teeth for surgery
Above The vampire bat (*Desmodus rotundus*) uses it protruding canine teeth to make blood-letting incisions.

Multipurpose teeth
Above The omnivorous bear has stabbing canines and numerous cheek teeth for both cutting and grinding.

Flesh-eating teeth
Above The lion has few cheek teeth, but their edges are razor-sharp and they slice easily through the prey's body.

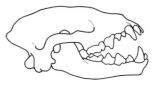

Bone-crushing teeth
Above The hyena has massive cheek teeth that can crush and grind all but the largest bones.

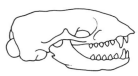

Grasping teeth
Above Fish-eaters, like the sea lion, have many recurved, pointed teeth that grasp slippery prey and hold it firmly.

we cannot be quite so precise when describing their hunting techniques as we can when dealing with the invertebrates. The latter are programmed to respond to given signals in a very precise way and rarely deviate from this instinctive programme.

Competition and specialization
Life is a continuous struggle for existence, as Charles Darwin pointed out over 100 years ago, and survival goes to the fittest. We have already seen this in respect of the prey animals, but it is equally true of the predators. There is tremendous competition for food, and only the fittest and most able manage to get enough. Such competition has led to a bewildering array of specializations, in both anatomy and behaviour, enabling their possessors to utilize virtually every kind of food. This is well seen in the great variety of

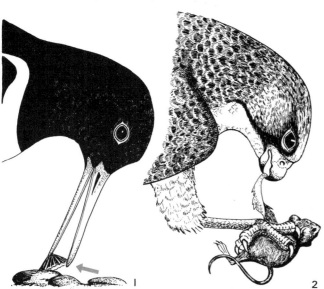

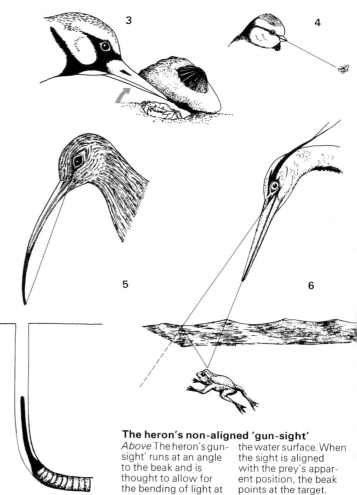

Beaks for many different diets
Above Birds' beaks usually have to take the place of both hands and teeth, and they vary enormously in shape and size according to the birds' diets. Many species actually have 'gunsights' on their faces, and it is thought that these help the birds to line up the tip of the beak with the prey. The oystercatcher (1) uses its stout, long beak as a pickaxe or crowbar to prise sedentary limpets and mussels from the rocks, and it has no need for 'sights'. The kestrel (2) does not need them either, for it catches its food with its talons and uses the hooked beak merely to tear up the prey after it has been caught. The turnstone (3) flicks stones over with its beak and uses its 'sights' to line up on the small creatures that it disturbs. The blue tit (4) also might benefit from its eye-stripes when sighting its small insect prey. The curlew (5) has a curved beak, but the 'sights' are aligned with the tip of the beak and assist the bird in probing the mud at just the right point to come up with a juicy lugworm or some similar prey. The curlew's eye can spot the burrows very easily.

The heron's non-aligned 'gun-sight'
Above The heron's gunsight' runs at an angle to the beak and is thought to allow for the bending of light at the water surface. When the sight is aligned with the prey's apparent position, the beak points at the target.

bird beaks, each adapted for a different diet. Even very small differences in structure can be important, for they may allow the animals to tap slightly different food sources. The penguins illustrate this point very well: the rockhopper and macaroni penguins breed in close association, but the macaroni is slightly larger and has a larger beak and it can thus take larger food. It can also dive to greater depths to feed, and so competition between the two species is reduced to a minimum.

The blue tit and the great tit show similar differences in feeding habits. Almost 90 percent of the insects eaten by blue tits are less than 5mm (0·2in) long, whereas less than 50 percent of those eaten by great tits fall into this category. Furthermore, when feeding in the trees, the blue tit hunts at an average height of about 9·5m (31ft) and the great tit at an average height of 7·8m (25·6ft). Separation of the two species becomes almost complete when we add David Lack's observations that the blue tit does 34 percent of its winter hunting among the twigs and buds of trees and only 7 percent on the ground, while the great tit does 50 percent of its winter hunting on the ground and only 5 percent among the twigs and buds. It is very doubtful that the two species could exist together without these differences in feeding habits.

Many much more elaborate speciali-

zations can be seen later in this book, all evolved to enable the predator to get the best possible deal from life. The long, herding tail of the thresher shark, and the remarkable tongue of the chameleon are just two examples. And it is a continuing story: just as the prey are getting fitter through the selective action of the predators, so the latter must get faster or more cunning if they are to continue to get their share of the food market.

Scavengers and predators

Scavengers, as we have already seen, are those animals that obtain the bulk of their food from dead plant or animal matter. They work hand in hand with the decomposers — the fungi and the bacteria — to form a sort of food chain in reverse. The scavengers eat up the dead material and incorporate some of its energy into their own systems, but they pass much of it out in a considerably finer form in their droppings. Smaller scavengers, such as the dung beetles and fly larvae, take the process further, and eventually the bacteria complete the cycle by breaking down the organic

Egyptian vulture – an opportunist feeder

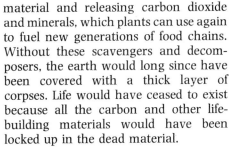

Below The Egyptian vulture is primarily a scavenger (see page 150), but it commonly eats eggs. It often drops them from a height to break them, but if they are too large to pick up, the bird will throw stones at them until they break, as shown in this picture.

material and releasing carbon dioxide and minerals, which plants can use again to fuel new generations of food chains. Without these scavengers and decomposers, the earth would long since have been covered with a thick layer of corpses. Life would have ceased to exist because all the carbon and other life-building materials would have been locked up in the dead material.

In global terms, the most important scavengers are the fly larvae and beetles, which can be found on every small bird and mammal within hours of its death. The larger and more spectacular scavengers, such as the vultures, the hyenas, and the jackals, are found only where there are large carcases to be dealt with. Beetles and fly larvae would take a very long time to dispose of a zebra carcase on their own — although they would manage it in the end — but a flock of vultures or a clan of hyenas can do the job in minutes, leaving just some pieces of skin and bone for the beetles to clear up. The hyenas do not even leave many bones, for they are able to crack and eat them with their massive teeth and jaws.

None of these large scavengers confines itself entirely to carrion; hyenas, long regarded as high priests of the scavenging life, actually get more food by killing than by scavenging in some parts of their range, and jackals also go in for a good deal of hunting. Vultures do not often kill, for they have rather weak talons, but they do sometimes jump the gun and attack weak and dying animals before they actually keel over, and then the vultures are technically predators. Conversely, many true predators will eat carrion when they get the chance — foxes, tigers, crocodiles, and gulls are all happy to supplement their catches with it, often when the meat is in an advanced state of decay. It is thus very difficult to draw a line between predators and scavengers, and the subject is even more complicated if we bring in the pirates that rob other animals of their prey. All we cay say with certainty is that these animals are all opportunists, able to take advantage of almost any kind of food that presents itself.

The parasites

A parasite may be defined as an animal or plant that lives in or on another species, remaining in close association with it and taking food from it without giving anything in return. Fleas, lice, tapeworms, and liver flukes are thus all parasites, but they are not really killers. They are generally very small in relation to their hosts and, unless they are present in very large numbers, they do not do much harm. Each parasite takes only a small amount of blood or other food from

Insidious invaders
Left The larvae of
Apanteles difficilis
have gradually eaten
their way through the
body of this fox moth
and reduced it to an
empty skin. The fully
grown parasitoid grubs
are now boring their
way out of the host's
skin ready to pupate.
In their early stages,
the *Apanteles* grubs
avoid eating the host's
essential organs, thus
ensuring that it stays
alive long enough for
them to complete
their growth.

Larders on spikes
Right This side-
blotched lizard was
caught by a logger-
head shrike and impaled on
a thorn in readiness
for feeding the chicks
in the nest.
Far right A rufous-
backed shrike with its
locust prey impaled
on a spine beside it.
Although prey is often
hung up near the nest
to feed the young, the
main reason for spiking
the prey is to hold it
firmly while the shrike,
also called the butcher-
bird, dismembers it
with its hooked beak.

its host, and the host can make good this loss without much trouble. It is obviously not in the interests of the parasite to kill its host, for in doing so it would destroy its own home and its continuing source of food, and would itself become extinct.

There is, however, one group of very specialized parasites that do kill their hosts – but not until they have finished with them. These parasites are insects and the best-known are the ichneumon flies (which are relatives of the wasps) and some species of true flies. They are collectively called parasitoids and their hosts are generally the young stages of various other insects.

The typical ichneumon life history begins when a mated female seeks out a caterpillar as a host for her offspring. She uses her antennae to smell it out, and cryptic coloration or camouflage, which conceals many insects from predatory birds, is no protection against her. Having found a suitable host, the ichneumon plunges her ovipositor into it and lays one or more eggs. The eggs hatch and the ichneumon grubs begin to nibble the tissues of the host, but they avoid the essential organs for the time being and they concentrate on the reserves of fat and the muscles. The caterpillar becomes rather lethargic, but it remains alive and continues to feed. The grubs inside it grow rapidly, and when they are nearly fully grown they attack the vital organs of the host and kill it. Pupation of the

ichneumon grubs may take place inside the caterpillar skin or outside it, and a new generation of adults emerges later.

Most people with a garden are familiar with the little yellow cocoons that cluster around the shrivelled skins of cabbage white caterpillars in the autumn, but few know just what they are and many squash them in the belief that they are 'caterpillar eggs'. In fact, they are the cocoons of an ichneumon relative called *Apanteles glomeratus*, and the insects have done us a good turn by destroying the cabbage pest. Left alone, the cocoons will yield a new generation of parasites to attack next year's caterpillars and so reduce crop losses.

These parasitoids are less dramatic in their action than the lion and other more familiar predators, but they are certainly no less efficient as killers. The only real difference is that, being smaller than their victims, they kill slowly and make one victim last for as long as they need it. The hunting wasps form a link between the parasitoids and the true predators. They paralyze caterpillars and other prey and store them for the young wasps to eat over a period of some months.

Filling the larder
Apart from the hunting wasps, which make provision for their young by stocking the nest with prey, few carnivorous animals make any attempt to store food. One of the reasons must be the difficulty

of keeping the flesh in a fresh condition – a problem that the wasps have solved by merely paralyzing the prey with the sting and not actually killing it. The moles have also hit upon the idea of paralysis to keep their stores of earthworms in good condition, but they go about it in a very different way: they bite the head end off each worm. This does not kill the worms, but it ensures that they cannot crawl away – at least for the time being. Drastic as it may seem, this treatment is not fatal to the worms, and if the moles do not return to their larders within a few weeks the worms regenerate their front ends and crawl away.

A number of cats, notably the leopard and tiger, store their leftovers from one day to another, the leopard being remarkable for its ability to carry quite large antelopes high into the trees. Crocodiles often conceal their leftovers at the bottom of the river, but none of these stores can really be regarded as a larder for the future because the food is generally eaten within two or three days. It is just that the animal cannot eat the whole of its prey at one sitting and is reluctant to leave it to the ever-present scavengers. The caches of the wolverine are true larders, however, and certainly help the animals through bad periods. The colder climate, with temperatures often well below freezing, makes storage much more feasible than it is in the tropics.

Most of the red fox's caches are certainly short-term reserves, and so, perhaps, are those of the weasel.

The most unusual food stores, however, are those of the shrikes, or butcher birds. Like the birds of prey, the shrikes catch their prey with their feet, but they usually kill it with a bite from the sharply-hooked beak. Insects, lizards, and small birds and mammals are the main foods, and they are very often impaled on thorns or even barbed wire to hold them firmly so that the shrikes can tear them to pieces. The prey is usually eaten immediately, but during the breeding season prey may be hung up in quantity near the nest and taken down as it is needed to feed the youngsters.

Second-hand food
Most predators kill their own food, but some animals rely on other species to do the work for them and then move in to take the spoils. Hyenas do this in some places, watching lions make a kill and then chasing the lions away. In other places the roles may be reversed, and it seems likely that such behaviour patterns have been learned in certain areas and passed down through the generations in particular clans or prides. This form of predation is known as piracy, and it is also well developed in a number of gulls and other seabirds, such as the skuas and the frigate-birds. Fish eagles and sea eagles are also adept at this form

of food-getting. The pirates chase other birds and force them to give up the food they are carrying. Young skuas and other pirates begin this behaviour almost as soon as they can fly and it is clearly instinctive. It may have developed from the habit of scavenging around the nests of other birds.

Man, of course, uses animals to catch food for him. Falconry is primarily a sport today, but it started as a method of catching game birds and others for food, with the falconer sending his peregrine streaking after a bird in just the same way as the modern wildfowler sends lead shot after it. Ferreters put their ferrets — domesticated polecats — into warrens to flush out rabbits, and fishermen in some parts of the world use cormorants to catch fish for them. The prize, however, must go to a small crab known as *Lybia tessellata*. This enterprising animal carries a sea anemone in each claw and thrusts them at fishes and other animals that come within reach. The stinging cells of the anemones ward off predators, but it seems fairly certain that they also disable a number of creatures and that the crab then uses these creatures as food.

Killing other than for food
Only man kills other animals with any regularity other than for food. Killing among wild animals is generally simply a matter of getting enough to eat, and

with hunger satisfied there is just no urge to kill. Any predator that went on killing just for the sake of it after filling its belly would very soon run out of prey, and any instincts in this direction are quickly eliminated. Species that have no significance as prey are generally ignored, although there may be some squabbling when two species meet on a common feeding ground. Hyenas, for example, may push vultures aside from a carcase. Lions may join in as well, and then there is sometimes a real fight, in which either lions or hyenas may be killed but not eaten.

Fighting most commonly occurs within a species, but here it has an entirely different basis: it is a trial of strength aimed at getting or retaining control of a territory or a group of females. We see such fights among lions, among seals at breeding time, and even among fishes and birds and such otherwise peaceful herbivores as the deer. Rarely do such fights end in death; generally the loser is not even injured. It is not in the interests of a species to kill off members of its own kind, especially among the herbivorous animals, for the weaker individuals, which lose the territorial fights and the rights to breed, form a valuable buffer between the breeding groups and the predators.

In conclusion, it is worth mentioning very briefly some of the animals that do kill without eating their victims. Man is the obvious example, but there are also some interesting ones among the bees. Honeybees throw out the males, or drones, at the end of the summer and sometimes sting them to death, for the males are of no further use and would only use up valuable food stores during the winter — so food is still involved, even if indirectly. Cuckoo bees of the genus *Psithyrus* lay their eggs in the nests of various bumble bees (*Bombus* spp) and rely on the bumble bee workers to rear them. The *Psithyrus* female usually kills the bumble bee queen so that no more bumble bee eggs are laid and the workers devote themselves entirely to rearing the cuckoo bee's youngsters.

Similar assassinations are carried out by a number of ant species. *Bothriomyrmex decapitans* is a well-named example from the Sahara. The mated queen allows herself to be dragged into the nest of a species of *Tapinoma*, and she then takes refuge in the royal chamber. She climbs onto the back of the *Tapinoma* queen and, using her large jaws, she decapitates the rightful queen and takes over as head of state. Her eggs and young are tended by the *Tapinoma* workers, but the latter eventually age and die and the whole nest is then populated by *Bothriomyrmex* for the rest of its existence.

The Ambushers

Everything must look normal, nothing must alert the
suspicions of the prey and cause it to turn back — patience and concealment ar
thus the ingredients of the successful ambush, and the viper lurking
under the sands of the desert knows this instinctively: few rodents
scampering over the desert floor would notice the viper until
it struck — and then it would be too late.

The Art of the Ambush

An ambush is a surprise attack on a victim by an enemy who has been lying in wait. This tactic has often been used in human warfare but, as with so many other human 'inventions', the ambush had been developed by animals long before man appeared on the scene. The ambush is used with great effect by a wide range of animals, both on the land and in the water. The warring human sets his ambush in a place through which the victim is likely to pass and, although they give no conscious thought to the siting of their ambushes, other animals do the same. Millions of years of evolution have ensured that the animals instinctively station themselves in suitable places; those that failed to select suitable sites have failed to survive.

Ambush or trap?

The victim of a true ambush is unaware of anything unusual until it is caught, but there are a number of examples in which the predator lures its victims into the ambush by means of bright colours or other attractions. For the purposes of this book, most such predators are regarded as trappers (see page 46). There is no clear-cut distinction between the ambush and the trap, however, and many animals fall into the 'grey area' between them. Some brightly coloured sea anemones, for example, may lure their prey to them, but their method of dealing with it is precisely the same as that employed by the less conspicuous *Hydra* and the jellyfishes. For convenience, all are treated as ambushers.

The trappers clearly include all those creatures, such as spiders, that actually make traps. But what of the floating *Pleurobrachia* or sea gooseberry? This delicate animal and its numerous relatives hang sticky nets in the surface waters of the sea and catch small prey in much the same way as spider webs trap flies, but the nets are actually parts of the living animals. Again, such creatures could be considered to be either ambushers or trappers, and they are treated as ambushers here simply because they do not construct anything in their external environment.

The role of camouflage

With surprise attack being the essential ingredient of an ambush, it is clearly in the interests of the ambushers that they should be invisible to their victims until the last moment. Camouflage therefore plays an important role in the lives of the ambush specialists. Many praying mantises, for example, are beautifully camouflaged as they sit on tree trunks or among leaves and wait for food to arrive. Their coloration clearly protects them from some of their enemies as well, but the mantises are really masquerading as wolves in sheeps' clothing and provide good examples of aggressive camouflage.

Some of the larger bush crickets, such

Deceptive predators
Left Ambushers come in all shapes and sizes. Despite their beautiful flower-like appearance, these rosy anemones, with their waving tentacles and batteries of stinging cells, are as deadly as any fanged predator to the small sea creatures which come within reach of their numerous grasping arms.

Hanging ambush
Right The scorpion fly *Bittacus* hangs by its front legs and grabs any passing small fly with its back legs. Each leg ends in a single hooked claw and the last two segments, which are toothed along their inner margins, can be snapped together like a pocket knife to give a secure grip on its struggling prey.

Camouflaged killer
Above The gaboon viper relies on disruptive camouflage for concealment. The different colours and patterns on its body break up its outline and merge it into the background of dead leaves on which it lies. This African snake is one of the puff adders, puffing air from its lung when it hisses with annoyance. Like all vipers, it has long poisonous fangs.

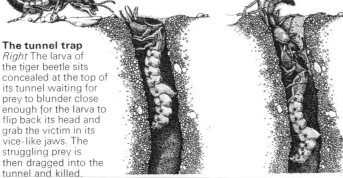

The tunnel trap
Right The larva of the tiger beetle sits concealed at the top of its tunnel waiting for prey to blunder close enough for the larva to flip back its head and grab the victim in its vice-like jaws. The struggling prey is then dragged into the tunnel and killed.

Detecting the approach

A predator lying in wait for food must obviously be able to detect the approach of suitable prey, and it must also be equipped to catch it efficiently when it comes within range. All the major senses are used in ambush situations, but sight and smell are especially important. Vibrations also play a major role in the detection of approaching prey, particularly by spiders and snakes. Most sea anemones and their relatives have to be touched before they fire off their stinging cells (see page 38), but some will fire when stimulated by scent particles in the water.

Capture must be swift and sure, for the ambusher is not equipped to chase after its victim if it does not catch it cleanly at the first attempt. Many ambushers, such as the tiger beetle larva and various snakes, use their powerful jaws to seize and hold their prey. Some ambushing spiders also strike with their jaws, but others first pounce with their strong front legs and then quickly bring the poison fangs into play to subdue and paralyze the victim.

Front legs are also used by the mantises (see page 30), but one of the strangest methods of capture must be that employed by the hanging scorpion flies of the genus *Bittacus*. These slender insects hang from the vegetation by their front legs and use their hind legs to capture small flies and other insects.

Arms with powerful suckers are used by the octopus (see page 196), while the sea anemones and their relatives use the batteries of stinging cells on their tentacles to catch and kill their prey. The sea gooseberries have no stinging cells, but their incredibly sticky 'lassoes' are quite up to the task of trapping and holding small fishes and other planktonic animals that come within range.

Hunting for food

Although the ambushers are well adapted for their rather sedentary way of life, they cannot be sure that food will come along at regular intervals. Most of them will get hungry from time to time, and if they get too hungry many of them will actually leave their stations and go hunting for food. Tiger beetle larvae certainly leave their burrows and go in search of food on occasion, and crab spiders are very often seen moving over the vegetation and pouncing on small insects. The octopus also goes hunting very regularly; in fact, it probably spends as much time in active hunting as it does sitting in ambush and, for the sake of convenience, it is dealt with in detail in the hunting section alongside the related squids and cuttlefishes — all marine molluscs with suckered tentacles.

as *Saga*, are similarly camouflaged while sitting among the leaves. Many of the crab spiders can actually change their colours to some extent to match the various flowers on which they wait for their victims. Trapdoor spiders conceal themselves inside their silken tunnels and peer out through the slightly open door as they wait for a victim to come within range.

Tiger beetle larvae ambush their prey in much the same way as the trapdoor spiders, but they construct less elaborate retreats. Each larva excavates a burrow, usually in sandy ground or in a steep bank, and uses its hard head to block the entrance. The burrow may be 30cm (12in) deep, but the larva anchors itself firmly at the top by means of its strange zig-zag shape and a number of stiff spines on the body surface. It is difficult to drag out an unwilling tiger beetle larva, but the insect can move very quickly when necessary. As soon as it

detects suitable insect prey within range, it flicks out its head and impales the victim with its huge jaws. It then retires to the bottom of the burrow to enjoy its meal. Cockroaches, caterpillars, earwigs, and most other crawling insects are readily taken by the tiger beetle larva, which probably detects them by means of vibrations and scent.

One insect knows how to get past the formidable jaws, however, and that is the female *Methocha* — a small, ant-like creature allied to the hunting wasps (see page 148). Perhaps her scent is not attractive to the young tiger beetle, but she certainly eludes its jaws and descends into the burrow. She then stings the soft abdomen, inducing rapid paralysis in the larva, and lays an egg on it. Having done this, *Methocha* emerges from the burrow, fills it with soil to conceal its position, and leaves her offspring to hatch and feed on the doomed young tiger beetle.

Mantises – The Living Gin Traps

The nearest thing to a gin trap in the animal kingdom must surely be the praying mantis, whose wickedly spined legs trap and hold its prey as securely as any man-made invention. The insect gets its name from the attitude in which it sits and waits for a meal to arrive, with the front legs neatly folded in front of the face, very much in the manner of a person at prayer. But this saintly attitude is merely a cover-up for a ruthless killer, and preying mantis would be a more appropriate name for the insect. There are about 2,000 species, most of them living in the tropical parts of the world, and they are generally simply called mantises. The 'original' praying mantis is the large European species known scientifically as *Mantis religiosa* and in France as la mante religieuse.

Patience rewarded

The majority of mantises are green or brown – many species have both green and brown forms – and they blend in very neatly with their vegetable backgrounds. Looking quite inoffensive, they sit motionless for long periods, gripping their supports with their last two pairs of legs and holding the first pair up in front of the face – but not far enough up to cover the huge eyes. These detect the slightest movement in the surroundings and the head moves round – sometimes very slowly and stealthily, sometimes rather rapidly – until the striking triangular face is looking straight at the cause of the disturbance. The mantis has a long, mobile pronotum, or neck, and a very flexible joint between this and the head, thus allowing the head to turn in almost any direction.

If the mantis sees an insect it may move stealthily towards it, but it usually remains where it is and watches intently, just waiting for the prospective victim to come within range of the ambush. The mantis moves its head round slowly to keep the eyes and front legs in line with the unfortunate insect, and it may adjust its position from time to time by carefully moving its four posterior legs to give itself a more comfortable stance. There may be a preliminary stretching of the front legs, perhaps to test the range or to make sure that everything is in working order, but the eyes never leave the prey. Then, in a fraction of a second, the legs shoot out and back and before the observer realizes what has happened the mantis is tucking into its meal.

The capture mechanism

The femur of the mantis front leg bears two rows of stout spines on its lower surface, and the tibia, which has two rows of smaller spines, folds back

Eaten alive by the praying mantis
Above A locust has been caught in the front legs of *Mantis religiosa*. The rows of spines on the grasping segments of the leg dig into the prey and hold it tight. The mantis then begins to devour the struggling prey. Wings and legs are occasionally dropped as the mantis cuts through their bases.

against it when the insect is at rest. The tibia also ends in a powerful claw. When the legs are fired out, the prey is often impaled on the tibial claws and the legs are immediately folded back to trap the victim between the spines.

The powerful jaws of the mantis come into play right away, biting deeply into the victim to quieten its struggles. The first bite is often taken from the thorax, which contains the nutritious flight muscles, but the mantis is not fussy about where it starts and the exact position depends largely on how the legs have grasped gin prey. Large chunks of flesh are carved out and chewed up small enough to pass down the mantis's very narrow neck. Nothing is wasted: even the hard head capsule of a grasshopper poses no problems for the mantis jaws, and the dry wings of moths are munched up in a way reminiscent of a rabbit munching lettuce.

Grasshoppers, butterflies and moths are staple foods for most adult mantises, but flies are also caught in large numbers, especially by the smaller mantises, and almost any insect will be snapped up if it comes within range. Warning colours mean nothing to mantises, and Fabre described how some enterprising mantises would take up station by the burrows of solitary wasps and wait for the wasps' return. The mantises often secured two meals for the price of one by this method, for the wasps often returned with caterpillars or other insects with which to stock their nests (see page 148).

Some mantises tackle prey larger than themselves, certain tropical species having been seen to take birds and lizards. The mantis abdomen swells visibly during a large meal, but there is a limit to its volume and extra-large meals are not always finished – and they are not finished later, for the mantis shows no interest in dead food.

Washing-up after a meal is as important for the mantis as it is for us – perhaps even more important – and the

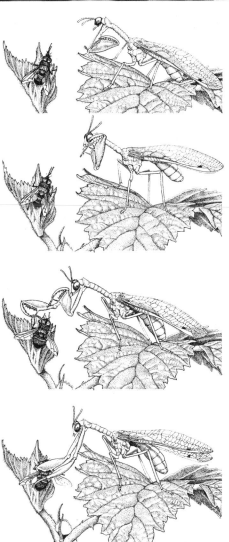

The look-alike trap
Left The mantis fly (*Mantispa*), seen here in the act of catching a small fly, is not related to the praying mantis, but it behaves in the same way and its insect-catching front legs are built to almost exactly the same plan.

The flower mantis
Above The young *Hymenopus coronatus* is pink and lobed and closely resembles the flowers on which it sits. It even moves with small, jerky motions so that it looks like a flower trembling in a slight breeze.

insect spends a considerable time cleaning its legs after eating. The jaws work their way up and down each front leg in meticulous fashion, picking off every minute piece of meat and other debris and keeping the spines in perfect condition. Clogged and dirty spines would clearly be less efficient at catching and holding the next meal.

Fatal mating

As well as eating the full range of other insects, mantises regularly eat their own kind, the best known examples of this cannibalism being the consumption of males by their mates during copulation. The normal mating position is with the male, who is somewhat smaller than the female, standing on her back and curling the tip of his abdomen round to meet hers. But the female often turns her head slightly, puts her arm around his neck, and calmly starts to bite his head off. Beheading does not impede the all-important business of sperm-transfer, however, and the female may eat virtually the whole of her mate's body before copulation is complete. The hapless male cannot even appreciate the

fact that his body is helping to nourish his progeny developing inside the female.

Flower mantises

Although the majority of mantises are cryptically coloured or camouflaged, there are a number of brightly coloured species in the tropical regions. Some hide in flowers and match the petal colours very closely. This is really no different from green mantises that sit in ambush on leaves, but the flower-inhabiting species do have the advantage that more prey insects are likely to visit the flowers than the leaves. The most striking mantises are undoubtedly those that use their shapes and colours to lure prey to them. According to our definitions of trappers and ambushers on page 00, these mantises are trappers, but it is convenient to mention them here with the rest of the mantises. They are brightly coloured and the legs and thorax have flat, petal-like extensions that make the whole insect look like a flower. Nectar-feeding insects are attracted to the flower mantises, but instead of nectar they find death in the vice-like grip of the mantises' front legs.

Crab Spiders - Beautiful but Deadly

The crab spiders get their name from the squat, crab-like appearance of most members of the group and also from the way in which they move — always sideways or diagonally forward, never walking straight to the front. But walking is something that the typical crab spider does not do very often. An individual may sit on the same leaf or flower for days on end, catching what insects come its way, sucking them dry, and discarding their empty husks before settling down to wait for the next victim. Some of the species pounce on their victims when they come within range, but the majority wait until the prey is right within the grasp of the strong front legs: one swift and economical movement then clasps the victim firmly to the spider's head.

A waiting game

Many crab spiders ambush their prey on flowers, but other species lie in wait on leaves and tree trunks and quite a number live on the ground and feed on the small creatures that wander about on the soil and in the turf.

When lying in wait, the crab spider's first one or two pairs of legs are usually held out in front of the head, while the other legs maintain a firm grip on the support. Approaching victims are detected by vibration and also by eye, and the spider smoothly adjusts its position

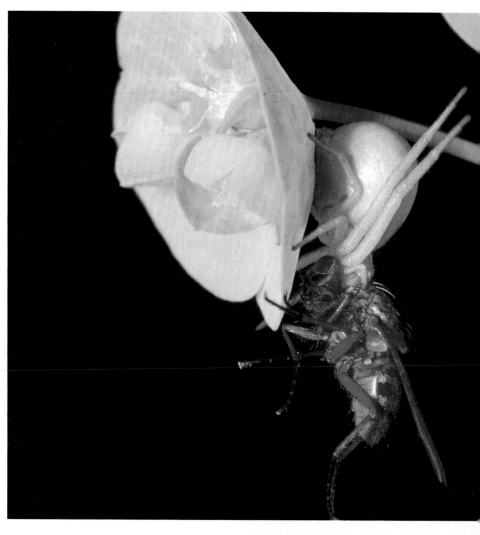

Danger in a flower
Above right The beautiful crab spider *Misumena* relies on its camouflage to conceal it amongst the flower heads. It remains motionless for hours on end just waiting for an unsuspecting insect to come within grasp.

Green as the grass
Left Micrommata virescens hangs upside down amongst grass leaves, the stripes along its back serving to break up its shape.

if necessary so that its four lower eyes — two of them often much larger than the others — are lined up on the prospective meal, following its every movement.

Whether the spider leaps on to its prey or merely grabs it when it gets close enough, the initial capture is with the front legs, but the spider's fangs are rapidly brought into use to inject a paralysing venom into the victim.

The spider's deadly venom

The fangs of the crab spider are almost always sunk into the victim just behind the head. This is, of course, the most accessible place when a bee or a fly lands in front of the spider and puts its head

down to feed, but it is also the most efficient place in which to inject the poison. The crab spider's venom is a nerve poison, producing rapid paralysis in the victim. This is clearly an advantage, for a struggling insect would only draw attention to itself and the spider as well. Total paralysis can be obtained most rapidly by injecting the venom into the main nerve cords of the neck, and this is just what the spider does. When a victim is caught and bitten in some other region, the spider usually turns it round and gives it a second deadly bite in the neck.

A wide range of prey

Some of the crab spiders attack quite large prey. The flower-haunting *Misumena vatia*, for example, regularly catches bees and butterflies, although the spider itself is only about 1cm (0·4in) long. Many an entomologist can recall stalking carefully up to a prize butterfly sitting on a flower, only to find the insect dead and firmly held in the grasp of white death — a popular name for the crab spider *Misumena*.

The prize for bravery (or stupidity), however, must go to a Ugandan crab spider, *Platythomisus insignis*, which was once found feeding on a mantis 8cm

Instant paralysis by the injection of poison
Above The crab spider *Xysticus* grabs an unsuspecting fly and sinks its fangs in the soft joint just behind the fly's head.

(3·2in) in length. The spider was a mere 1·9cm (0·75in) long. Such a case is exceptional, but it does show that the crab spiders are not worried by large size. There is, however, a lower size limit for prey, below which the spiders rarely bother to attack. The studies of W. S. Bristowe suggest that few spiders bother with prey less than about one sixth of their own length.

The flower-inhabiting crab spiders, such as *Misumena vatia*, obviously feed largely on flying insects, and some have

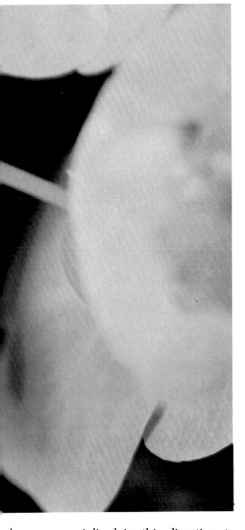

become specialized in this direction to the extent that they ignore crawling insects. Conversely, the ground-dwelling crab spiders feed on crawling insects and ignore or even shy away from flying species. Several crab spiders specialize in ant-eating, and some have evolved remarkable similarities to their prey. This 'wolf-in-sheep's-clothing' approach may facilitate prey capture in a few species, but the resemblance to ants is much more likely to protect the spiders from attack by birds and lizards. Most ants rely on scent and touch to explore their surroundings and they would not be fooled by a spider that simply looks but does not smell or react like an ant.

Masters of camouflage

Crab spiders, particularly those that sit on flowers and leaves, are masters of camouflage. They blend in beautifully with their backgrounds and, because they sit so still, they are very easily overlooked. Experimental work has indicated that such concealment is of great value in obtaining food. Bristowe and various other workers have experimented with coloured beads and pebbles placed on dandelions and other yellow flowers. In one experiment ten times more bees and hoverflies visited the

flowers with yellow pebbles on them than visited flowers with black pebbles, and many of the insects were visibly repelled by flowers with black pebbles on them. Clearly, the crab spider is more likely to get a meal if it is the same colour as the flower, and the spiders 'know' this: put one on a flower of the wrong colour and it will very soon crawl away and search for one of the right colour.

Misumena vatia and some other crab spiders can change their colours over a limited range to match different flowers. *M. vatia* ranges from white – hence 'white death' – to deep yellow, and if you take a white one and confine it in a yellow environment it will take on a distinct yellow colour within a day or two. The preponderance of white and yellow flowers in spring and early summer gives the spider a good chance of finding a suitable flower without much trouble.

One of the best examples of crab-spider camouflage is shown by *Phryna-rachne decipiens*, which sits on leaves and resembles bird droppings. The spider is black and white, and it sits on a small white web of silk which even has a 'blob' at the lower end to complete the resemblance. Butterflies and other insects

Camouflage – the vital subterfuge

Above This crab spider relies both on colour and shape to remain hidden. Its dull brown-grey colour resembles that of the bark on which it sits. Its body is flattened and is pressed closely to the bark to avoid throwing a revealing shadow. As long as it remains motionless, there is little chance of its disguise being penetrated.

regularly feed at bird droppings, to which they are attracted by the salty constituents, and many make the mistake of landing on *Phrynarachne*. They never get up. *Phrynarachne* may have evolved this clever scheme primarily in connection with feeding, but it is perhaps more likely that the camouflage developed initially as a protective device, for no bird is going to take much interest in its fellows' droppings.

Although most crab spiders are rather squat, there are a few long-bodied members of the group that inhabit clumps of grass. *Micrommata virescens* and the various *Tibellus* species are good examples. They sit on the grasses with their heads pointing downwards and their legs held fore and aft along the grass blades, perfectly camouflaged by their brown or green colours. Clumps of special hairs on the feet enable the spiders to move safely on the vertical surfaces, clambering about with ease.

The Ingenious Trapdoor Spiders

Imagine walking along the street and suddenly being grabbed by the ankles and dragged into a partly-open manhole, and you can imagine the fate that befalls the victims of the trapdoor spiders. These masters of civil engineering live in the warmer parts of the world and, although they are not all strictly ambushers, they construct some wonderful retreats from which they shoot out to grab their prey.

The structure of the burrow

Like the tiger beetle larvae, trapdoor spiders usually excavate their tunnels in the soil, although a few species construct their homes in tree trunks and bark crevices. Sloping banks are frequently chosen by the ground-dwelling species, and the tunnels are excavated with the enormous, downward-pointing fangs. (Most spiders have horizontal fangs.) The basal parts of the fangs are equipped with strong spines which act like rakes to sweep away the excavated soil. The walls of the burrows are lined with silk as the spider digs, and the finished burrow may be as much as 30cm (12in) deep and 3cm (1·2in) in diameter.

The lid, or trapdoor, is made at an early stage and it consists of silk mixed with a greater or lesser amount of soil. The silk hinge extends along one side for about a third of the circumference and allows the lid to fall into place under its own weight. Mosses and other plant material are often gathered and 'planted' on the door, providing perfect camouflage. This camouflage provides good protection against spider-hunting wasps (see page 148), while the door itself also fits snugly enough to prevent flooding. The trapdoor spider's burrow has rightly been called a triumph of protective architecture, being safe and comfortable.

The typical burrow is a single tube and, in its most familiar form, has a thick trapdoor that fits into the neck just like a cork fits into a bottle neck.

When the lid is closed it is almost impossible for an enemy to enter, even if it finds the burrow, for the spider itself acts as a living bolt, anchoring its legs to the walls of the burrow and digging its fangs firmly into the underside of the trapdoor. One set of experiments with a spring balance indicated that the average spider – itself weighing little over 3gms (0·1oz) – resisted with a pull equivalent to 123gms (4·3oz).

Some species make very thin doors – known as wafer doors in contrast to the cork-type already described – with broad flanges that overlap the edges of the burrows. Although it may be well camouflaged, this type of door gives little protection against a determined enemy and several of the trapdoor spiders with this kind of door make a second and much stronger door a bit further down

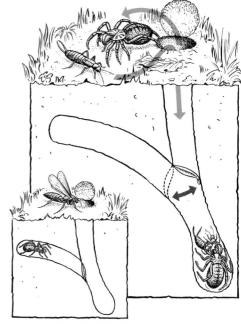

The sinister spider
Below The funnel spider *Atrax formidabilis* with its eight eyes and massive stabbing jaws must be a walking nightmare to any small invertebrate. The first pair of small leg-like appendages are called palps and are important tactile sense organs used to feel the prey and locate its soft vulnerable joints. Each of the eight, jointed, walking legs ends in two claws. Each joint moves in a different direction to give complete mobility to the leg. The hairs covering the legs are sensitive to vibrations and special sensory patches are present which react to heat, pressure, humidity and chemical odours. Sight is important in the last stage of the location of the prey, pinpointing its position as a target to pounce on.

Safety and danger
Above right Many trapdoor spiders have branching burrows with doors closing off either the branch or the main tunnel. If a dangerous predator like the hunting wasp enters the tunnel the spider can lock itself safely out of reach in the branch. But if prey blunders in, it can shoot out, closing the extra door to block the victim's escape.

Threat from below
Right Ummidia nidicoleus uses its powerful fanged jaws to dig out its subterranean burrow. The sides are cemented with silk. Finally a trapdoor is placed on top, well camouflaged with soil and small pieces of vegetation. Lurking safely at the mouth of the tunnel, the spider lunges at any prey passing within range.

the burrow. This second door is sometimes associated with a branch burrow into which the spider can retreat if attacked: the door is hinged in such a way that it can close off either the branch burrow or the main shaft.

Method of attack

Having described their retreats in some detail, it is now time to look at the feeding habits of the trapdoor spiders, and we find an interesting range of variation according to habitat.

Spiders living in relatively moist places tend to be true ambushers, for plenty of small animals are likely to walk past the door. The daytime is normally spent resting in the burrow, with the door firmly shut, but at night the spider pushes up the door and sits with its head and front two pairs of legs

poking out from the partly open door. When an animal comes within range, the spider lunges forward with its front legs and palps and strikes with its fangs. The hind legs remain anchored in the burrow, allowing the spider to retreat rapidly with its victim and close the door.

All kinds of insects may be taken, and in some places trapdoor spiders, which may number several to the square metre, take a heavy toll on locust hoppers. Lizards, birds, and small mammals are also caught from time to time, but this is not surprising when one realises that trapdoor spiders are close relatives of the big bird-eating spiders of the Amazon. Not everything that is caught is eaten, however, and quite a few unpalatable insects are violently ejected from the burrow almost as soon as they have been caught and tasted.

Detecting the prey

The hairy legs of the waiting spider are planted firmly on the ground and they certainly feel the vibrations of an approaching victim. This may be the main method by which some spiders detect their prey, but the eyes also play a part in sensing movement, especially in those species that live in more open environments. A study of various Australian trapdoor spiders revealed that those living among leaf litter have small eyes and poor sight, and therefore probably rely mainly on vibrations to tell of approaching prey. They would not in any case be able to see prey at any distance among the dead leaves. The trapdoor spiders living in more open ground, however, can see further and it is significant that their eyes are larger than those of the litter-inhabiting species.

Beyond the trapdoor

As already stated, trapdoor spiders are not all true ambushers: the group includes a number of species that emerge from their burrows to hunt on the surface. These hunters normally live in rather dry places where small animals are less abundant, and there is an interesting series linking them with the true ambushers. This series includes a number of species whose trapdoors are surrounded by radiating trip-wires made of silk. When an animal stumbles on a trip-wire the spider rushes out to bite it and drag it back to the burrow. This system enables the spider to cover a wider area than it could by simply sitting in its burrow.

Some Australian species actually collect slender twigs and arrange them radially around the trapdoors. The spiders maintain contact with the twigs by means of silken threads and they know immediately when one of the twigs is disturbed.

Spiders that rush out from their retreats normally make wafer-type lids that are thrown open and stay open without support when the spiders leave. The spiders have longer legs and much better eyesight than those that stay in their burrows, making them very adept at running after their prey.

A number of trapdoor spiders, including some living in southern Europe, actually make small webs just outside their burrows each night. The webs stand vertically like miniature tennis nets and, although only some 15cm (6in) long and 2 or 3cm (1·2in) high, they are remarkably efficient traps. The heavy trapdoor is held open by a few silk threads while the webs are being spun, and they remain open when the spiders retire to their burrows after spinning. When a victim is caught, the spider rushes out and bites it and carries it back to the burrow. The web is removed before dawn, whether it has been successful or not, and the trapdoor firmly closed. It seems probable that the old webs are added to the cork doors — an excellent example of recycling in nature, for the protective door can never be too thick for the spider.

The purse-web spider

Atypus affinis is a close relative of the trapdoor spiders and, being the only member of the group found in the British Isles, it is sometimes called the British trapdoor spider, although it does not actually make a trapdoor. Its retreat is a completely closed silken tube up to 30cm (12in) long and about as thick as a finger. Most of the tube is concealed in a burrow on sloping ground, but a small part is exposed on the surface.

When an insect walks over the exposed surface of the tube, the spider rushes up from the lower end and stabs its fangs right through the tube and into the victim. The teeth on the lower side of the fangs are then used to cut small slits in the silk so that the victim can be pulled into the tube. The cuts are quickly repaired, and the spider then retires to the bottom of the tube to enjoy its meal.

The purse-web spider is another of the many animals that belong to the 'grey area' between the ambushers and the trappers. It makes a web — the silken tube — which tells the spider that food has arrived, but the web does not actually catch the prey and the spider is thus best considered to be an ambusher rather than a trapper.

Purse-web spider
Right Atypus lives in a silken tube, most of which is below ground, but a small part extends along the surface. Bits of earth are stuck on the silk and woven into the web to strengthen it. When a small animal walks over the tube, *Atypus* (left) rushes out and stabs it through the roof of the tube.

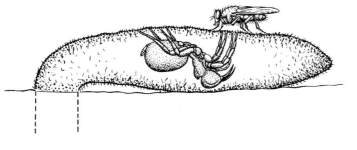

Trip-wire array
Below This Malaysian spider *Lyphistius batuensis* spends most of its time below ground, but silk trip-wires radiating from the trapdoor alert the spider to likely prey.

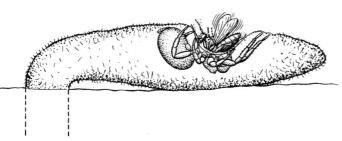

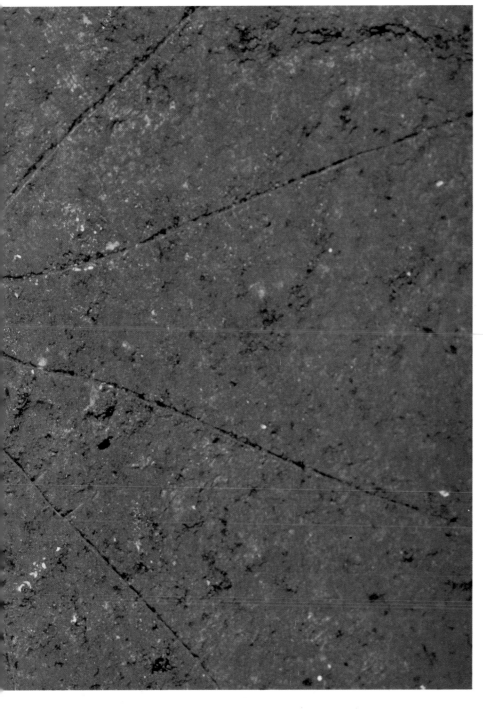

The spider's venom

A spider's poison glands are generally situated in the basal portion of the fangs or chelicerae, although some spiders, such as *Scytodes* species, carry the glands in the front part of the body. Ducts from the glands run through the chelicerae and open at or near the tip. Poison is thus pumped straight into the prey when the fangs strike home.

The fangs themselves are of two main types. Those of the trapdoor spiders and the large bird-eating species work in a vertical plane and are plunged into the prey more or less parallel to each other like two daggers. The fangs of other spiders work horizontally and close on each other rather like a pair of pincers.

The venom originally had a purely digestive function, and still contains digestive enzymes, but the spiders gradually evolved paralyzing agents as well. Most research has been carried out on the venom of the notorious black widow spider (*Latrodectus mactans*). This appears to have four main active constituents, three of which are active against insect prey. One of these three produces a quick knockout, while the other two are slower-acting poisons that destroy the prey's nervous system. The fourth component of the venom is active against mammals and birds, where it damages nerve endings in the muscles and produces paralysis and severe pain. Most spider venoms probably have a similar complement of active ingredients, although they differ in toxicity. The black widow spider is much feared on account of its painful bite, but it is rarely fatal. The Australian funnel-web spider (*Atrax robustus*) is much more dangerous.

The brown recluse spiders (*Loxoscelis*) of America have hemolytic factors in their venoms. These break down the general tissues of mammals and cause extensive damage around the initial fang punctures. The painful nerve poisons and hemolytic components are not necessary for killing prey, and have probably evolved as a protection against vertebrate enemies that may threaten them.

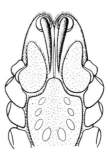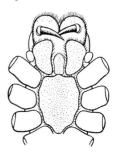

Two contrasting designs of spider fangs
Above Atypus has a pair of massive dagger-like fangs orientated vertically for stabbing through its tube-like web into the prey.

Above Amaurobius has the typical pincer-like orientation of spider fangs, crossing at their tips. These are sunk in the prey's legs.

Hydra and Sea Anemones

Hydra is a small green or brown animal with a body no more than about 3cm (1.2in) long and rarely much thicker than a pin. There are several species, all living in fresh water. One end of the body is fixed to a stone or a water plant, while the free end bears a number of tentacles surrounding the mouth. These tentacles may be longer or shorter than the body. The creature can be found by searching aquatic vegetation, especially the undersides of water lily leaves, but it is not easy to see because its muscular body contracts to a little blob of jelly when disturbed by vibrations in the water.

Pin-pricks of death

One's first impression of this little soft-bodied creature, more like a plant than an animal, is not one of a ruthless killer, but to water fleas and other small aquatic creatures it is a deadly trap. Its deadliness is due to its minute stinging cells called nematocysts. Similar structures can be found in a few protozoan animals, but otherwise these weapons are confined to the coelenterates — the group of animals to which the hydra and the sea anemones belong. Large numbers of stinging cells occur on the tentacles, with smaller numbers located elsewhere on the body.

There are actually several different types of nematocyst, but they all have the same basic structure. The typical nematocyst is a rounded or egg-shaped body, rarely more than 1 mm (0.04in) across and embedded in one of the hydra's surface cells. It has a small cap and a 'trigger' called a cnidocil. Coiled up inside the nematocyst is a thread of variable length, inverted like the finger of a glove which has been pulled down into the palm section. The nematocyst is most commonly fired when a small animal blunders into a tentacle and touches the cnidocil, but it may also be fired simply in response to the scent of an animal in the immediate vicinity. In nature, it is probably fired in response to a mixture of chemical and tactile signals.

During firing, pressure builds up very rapidly in the capsule, forcing the cap off and everting the thread in an instant. The typical stinging nematocyst thread ends in a microscopic needle, which penetrates the water flea or other victim and at the same time injects a minute amount of poison. Several nematocysts are generally fired off together, and the prey is quickly paralyzed.

The nematocysts can be used only once, and when they have been fired they must be replaced. New ones are always being produced from plain

A safe haven
Right The colourful clown fish is immune to the stinging tentacles of the giant anemone. Here the fish can pick up any morsels of food dropped by its host and seek refuge from enemies.

Colourful killers
Right These beautiful stone flowers are, in fact, small animals called cup corals. Each individual animal sits in a limestone cup and is armed with a ring of tentacles covered in stinging cells that capture and paralyze its prey.

Feeding hydra
Left Unlike some of the colourful anemones the diminutive hydra is usually green or brown in colour and hangs concealed amongst the aquatic vegetation, waiting to grab and paralyze its prey. The hydra chemically senses the presence of its prey.

Armoured cells
Below When stimulated, the wall of the nematocyst capsule swells forcing off the cap and exposing the inverted thread. On contact with water the wall of the thread swells, lengthens and everts, pulling out more thread. In a flash the whole barbed filament is everted and penetrates the prey.

A paralyzing trap
Centre left Hydra has a bag-like body made up of two cell layers. At the top end is a mouth surrounded by hollow tentacles. Batteries of armoured cells are arranged along the tentacles. These shoot out threads that penetrate the prey and inject poison into it. This immediately immobilizes the prey.

Looping hydra
Left Hydra can move around in search of a suitable fishing ground by using the sticky nematocysts that cover the tentacles. It bends over and glues down its tentacles. Then, releasing its foot, it contracts down, swinging its body over its head to search for another foothold. The tentacles then release.

The stomach bag
Centre and below The paralyzed prey is pushed through the mouth by the waving tentacles into the bag-like stomach. Digestive juices are secreted into the stomach to break down the prey's tissues. Small food particles are engulfed by some of the amoeboid cells lining the stomach.

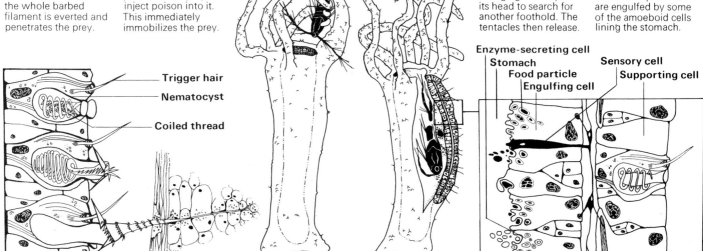

Trigger hair
Nematocyst
Coiled thread

Enzyme-secreting cell
Stomach
Food particle
Engulfing cell
Sensory cell
Supporting cell

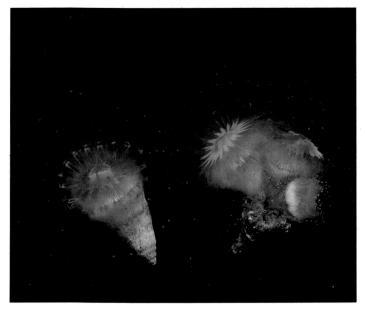

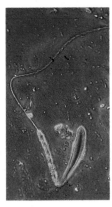

Coral nematocyst
Above The hollow everted thread of a discharged nematocyst acts like a hypodermic needle, penetrating the tissues of its prey and injecting venom.

'packing' cells in the main part of the body. They move into the body cavity and gradually work their way to the tentacles or any other site where new nematocysts are needed.

Not all nematocysts are of the stinging type. Some simply fire out sticky threads that entangle the legs of the victims, while others have barbed or coiling threads. For this reason, many biologists prefer to call the nematocysts thread cells instead of stinging cells.

Digesting the meal

Hydras clearly rely mainly on the prey blundering into the tentacles, but this is not quite as passive as it might seem. Chemical stimulation of the tentacles by a swarm of water fleas may cause the tentacles to sway this way and that, thus giving them a better chance of making contact with prey. When contact has been made and the thread cells have been fired, the tentacles slowly bend over towards the mouth and the captured prey is pushed into the mouth.

Preliminary digestion takes place in the fluids of the simple body cavity, but then cells lining the cavity actually engulf small particles in the manner of an amoeba (see page 16) and digestion is completed inside these cells. Indigestible matter passes out through the mouth, for there is no other body opening such as is found in most higher animals.

The various species of hydra get their colours from the green or brown algae that live in their tissues. In times of food shortage these tiny plants may be digested, but if a hydra fails to catch food for a while it will probably move to a new site. It moves in a cartwheel fashion, bending over so that special anchoring thread cells can be shot out to hold the tentacles to the substrate while the base is released to seek another attachment.

The sea anemones

The sea anemones are marine relatives of the hydra, but they are much larger and more colourful. One can be forgiven for thinking that they are plants when

they are seen in rock pools with their rosettes of tentacles spread widely or drooping gracefully from the rim. When uncovered by the tide, however, they contract into shapeless blobs of jelly.

Sea anemones have a rather more complex anatomy than the hydra, although the body still has a single cavity and a single opening. There may be only six tentacles, but there are usually dozens, and this means vast numbers of nematocysts. These are structurally identical to those of the hydra, but the venom varies from species to species and the sheer volume produced by so many nematocysts makes the sea anemones death traps even for quite large fishes.

Some sea anemones, including several species inhabiting the North Atlantic and the Mediterranean Sea, can affect human bathers who brush against them. The most common effect is similar to that of a stinging nettle, but there may be severe blistering of the affected area together with high temperature, partial paralysis and shock reaction.

Special protection

The sea anemones' powerful armouries protect them from most predators, but some fishes seem to be immune to the stings. The clown fish, for example, lives among the tentacles of a giant sea anemone which may be 1m (39in) across, and it seems that the slime on the clown fish's scales inhibits the action of the nematocysts.

Some fishes with horny mouths eat sea anemones without ill-effect, but the most unusual predators are surely some of the sea slugs. These beautiful creatures nibble the sea anemones and actually swallow the nematocysts without discharging them. The nematocysts then work their way through the sea slug tissues and take up position in the skin of the dorsal surface, where they serve to protect the sea slugs from inquisitive fishes that ignore their warning colours.

The corals

The corals belong to the same major group of coelenterates as the sea anemones, but most of them are much smaller. Each individual, known as a polyp, secretes a limestone cup around itself. Some corals are solitary, but the best known species are colonial: thousands of polyps, all descendants of one branching individual, are linked together in a single mass of limestone. Many such colonies make up a coral reef. The corals are armed with nematocysts just like the sea anemones, but they feed on smaller prey as a rule. The polyps are usually withdrawn into their cups by day, but they emerge at night and capture myriads of planktonic creatures.

The Science of the Trap

Apart from man, very few animals actually trap their food in the sense that they construct a snare or use some other external device to capture their prey. In fact, the only group of animals among which trapping plays a major role are the spiders, whose silken webs are familiar objects almost everywhere on earth. Even then, not all spiders make webs, but those that do have evolved a wonderful range of webs designed to catch a wide variety of flying and crawling insects. The construction of some of these flimsy, but nevertheless very efficient snares is described in the following pages.

With a few exceptions, such as the bolas spider (page 55), which may smell like a flower, the spiders do not lure their victims into their traps, but there are some other animals that certainly do attract their victims and lure them to their deaths. The New Zealand glow-worm (page 57) has a luminous lure as well as a sticky trap, but most of those animals that employ lures have no external trap – the victims are simply lured within reach of the predators' jaws and snapped up. Although no actual trap is constructed, it is convenient to consider these predators as trappers because the lure does bring in the prey and the predators do better than they would if they sat in simple ambush (see page 28).

The trappers, like the ambushers, require no great development of the sense organs to seek out their prey, nor do they need the speed of the hunters, although many spiders move remarkably quickly over their webs when a victim is trapped. The essential requirement is to know when food has arrived, and the trappers are usually aware of this through vibrations of the web or through the sight of small prey moving around the lure. Large jaws, with or without some kind of venom, are then the normal weapons for despatching the victim.

The ant-lion's pit

One of the simplest forms of trap is the pitfall. Men have been making pitfall traps to capture food and enemies for thousands of years, but the technique was in use by the ant-lion long before man appeared on the scene.

The adult ant-lion is a long-winged insect resembling a dragonfly, but the young insect is very different from the dragonfly nymph (page 178). It lives on land and possesses an enormous pair of jaws. Several of the species roam the ground as true hunters, but some construct pits in sandy soil. The pit is a conical depression, up to 10cm (4in) in diameter at the top, and it is dug by the

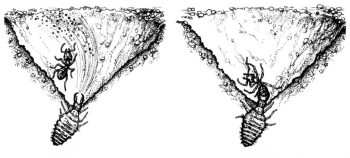

The pitfall trap
Left and right The ant-lion larva buries its soft vulnerable body in the sand at the bottom of a funnel-shaped trap so that only its head and pincer-like jaws are exposed. Any ant wandering into the rim of the trap usually starts an avalanche of loose sand and falls into the waiting jaws.

insect using its flat head and jaws as a shovel. It buries itself in the sand first, and then throws out the sand above it. When the excavation is complete, the ant-lion remains buried apart from the immense jaws, which poke out from the bottom of the depression like a pair of calipers. These clamp round the body of any ant that blunders into the trap.

The insect remains almost motionless, although it does move round on warm days to keep its body under the shadier and cooler side of the pit. Ants and other small insects that wander over the ground frequently tumble into the ant-

lion's pit, and they are usually unable to climb up the loose sides. If they do try to escape, the ant-lion throws sand at them to bring them back down to the bottom where the gaping jaws are waiting to inject poison and digestive juices.

The internal tissues of the victim are dissolved within a few hours, and the resulting solution is then sucked up through the ant-lion's channelled jaws as if through a drinking straw. The victim's empty skin is then thrown out of the pit, together with any excess sand that fell in with it during the struggle.

The silken trap
Left The success of the silken web as a trap for unwary insects can be seen in this photograph. These chironomid midges, which swam together, have blundered into the sticky threads and become firmly entangled in the web.

The pitcher trap
Above The ends of the leaves of this tropical plant are modified into pitcher traps. Small insects entering the rim slip on the waxy scales lining the walls and tumble into the digestive broth filling the lower third of the pitcher.

The fishing net
Below This caddis fly, *Plectrocnemia*, is found in gentle flowing streams, so it can build a silken net amongst the pond weed without danger of it being torn to shreds. The current of water carries small creatures down the funnel to the larva.

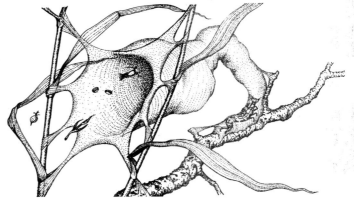

Fishing nets

Most young caddis flies construct portable cases around themselves and roam freely in the water, picking up whatever plant or animal matter they can find as food. A number of species, however, spin silken nets and use these to catch their food, although the nets are not moved through the water like trawls. They are fixed to the vegetation and they act more like drift nets, trapping whatever is brought along by the current. Several of the nets are conical, with the caddis larvae concealed within ready to pick off the food particles as soon as they are caught. Much of the material caught in this way is plant debris, but small insects and crustaceans are caught as well, and the larva usually eats them all.

Clever chimps

It used to be accepted that man was the only animal that made and used tools of any kind – in fact, tool-making was one of the criteria which anthropologists used to distinguish true prehistoric men from the various kinds of apes. This idea

has had to be somewhat revised, however, in the light of Jane van Lawick-Goodall's detailed work with chimpanzees.

The chimps may not actually *make* tools, but they certainly use modified natural objects. Leaves, for example, are crushed up and used as sponges to soak up drinking water, while small twigs are used to catch termites. The twigs are pushed into the termite mounds, where they are immediately 'attacked' by the insects, and then pulled out with several termites clinging tightly to them. The twigs are therefore simple forms of traps, and the chimps choose them carefully so that they are of just the right thickness. They also strip the leaves off when necessary, thus modifying and improving the 'tool' even if they do not make it in the first place.

A few other animals, such as the sea otter (page 200) and one species of Darwin's finches, use tools but they do not fashion them in any way and cannot be described as tool-makers in the sense that the chimpanzees are.

The Orb Web Spiders

The best known of the animal traps are undoubtedly the 'typical' spider webs – the wheel-shaped orb webs that adorn shrubs, fences, and most other outdoor objects in the autumn. The webs are there at other times as well, but they are generally smaller, and they certainly show up better when laden with dew on an autumn morning. They are made entirely with silk from the spider's body.

The orb webs are made by spiders of the family Argiopidae, which is distributed all over the world. There are certain differences in constructional details – some webs are generally vertical, while others tend to be slung at an angle, for example – but all orb webs consist basically of a number of radii, or 'spokes', on which the spider lays a spiral of sticky silk to trap flying insects.

The size of the web depends a good deal on the size of the spider; young spiders, for example, make smaller webs than adults of the same species, while small species generally make smaller webs than larger species. This is a very logical arrangement in view of the fact that large spiders need more food than small ones to keep active and healthy.

Constructing the web

Web-making is an instinctive act and does not have to be learned by the young spider – most young spiders are orphans when they come into the world and have no opportunity to learn anyway – and it is amazing how quickly and how accurately the webs are made. The sticky threads are attached to the radii to give exactly the right distance between each turn of the spiral. The spider uses its legs to measure everything and, during the course of making an average-sized web – say 25cm (10in) in diameter – an ordinary garden spider will walk, run, swing, or climb a considerable distance.

One typical web in the writer's garden had 17 radii, averaging 13cm (5in) in length, 30 turns of the sticky spiral, and at least 7m (23ft) of supporting threads attaching the web to the surrounding plants. This gives a total of about 37m (121ft) of silk, and quite a lot more was employed as scaffolding and then removed. But even this does not approach the full distance that the spider has to travel during construction, for it moves up and down the radii and supporting threads many times as it secures everything before starting the spiral – and yet the whole process can be completed within an hour.

Many webs, especially in tropical regions, are much larger than those of the garden spider. Species of *Nephila*, found in Australia and Southeast Asia, regularly make webs 2m (6.5ft) or more in diameter. Their thick, yellowish silk

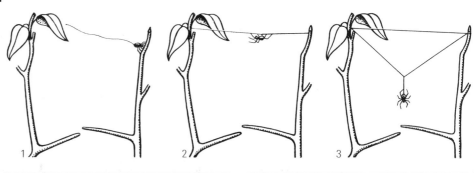

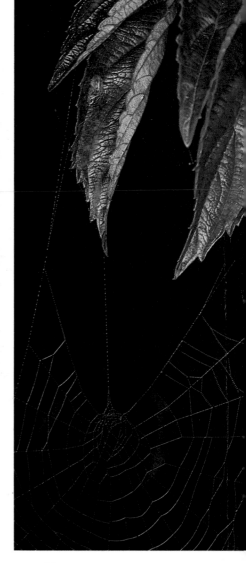

Parcelling the prey
Above Argiope aurantia is an orb-web spinner which has two distinctive silvery zig-zags one above and the other below the hub. This spider usually sits in the centre of its web waiting for prey like this dragonfly.

The open line
Right Zygiella spins an orb web with a missing segment. A stout line attached to the centre of the web passes to the spider's lair in a safe crevice. The spider usually crouches in its lair with a leg in contact with this wire.

is so strong that local fishermen use it for their nets. Each female spider can produce more than 300m (984ft) of silk every day – tremendous productivity for a creature only about 22mm (0.86in) long. The male, incidentally, is barely 5mm (0.2in) long and he often rides on his mate's back.

In addition to the usual butterflies, moths and other insects, the large *Nephila* webs often catch birds, but the spiders are perfectly able to deal with this source of food as well without risk.

Trapping and wrapping

When the web is complete, the orb spider either sits on a little platform in the centre or else moves to a sheltered retreat a short distance away from the web. The web is a very efficient trap and insects are so numerous that the spider rarely has to wait long for its food to be caught. Even when hiding in its retreat, the spider maintains constant contact with the web by holding on to one or more threads. As soon as it feels the vibrations of struggling prey, it zooms down to the middle and, by touching the various radii, detects exactly where the victim has been trapped. The spider does not seem to use its eyes for this at all.

Depending on whether the vibrations are large or small – indicating large or small prey – the spider advances slowly or rapidly and picks its moment to dart in to deliver a paralyzing bite. After this

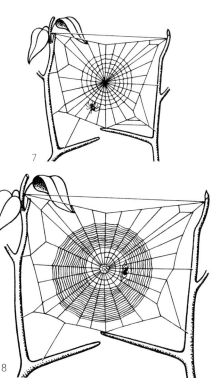

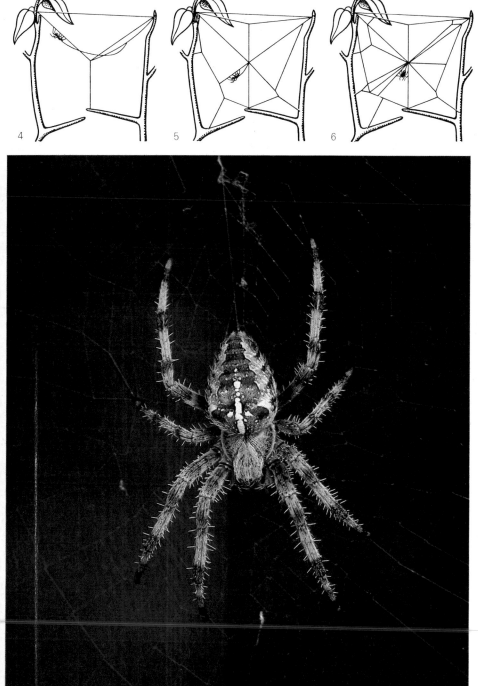

Crucial stages in spinning an orb web

1 The spider secretes a bridging-line across a suitable gap.
2 Pulling itself across the line, the spider eats it and replaces it with a permanent line.
3 Crossing back again it trails out another line to be anchored near the permanent line. Halfway along this line the spider drops.
4 Pulling the slack taut, a Y is formed. A thread is spun along one of the top arms, a line anchored to its centre and, crossing to the other arm, the thread is pulled taut, filling in the top triangle with a radial spoke.
5 This manoeuvre is repeated to fill in the other two triangles.
6 Starting from the centre, the spider measures out and fills in each radius.
7 Working outwards, a spiral scaffolding thread is spun.
8 Finally, working inwards the sticky threads are laid down.

The garden spider
Left Araneus diadematus often sits in the hub of its circular orb web its feet spread to sense any vibrations.

The spider's foot
Right The foot of a web-building spider has a hooked claw and two toothed claws that guide silk into it.

the prey is usually wrapped in a silken shroud and carried to the centre of the web or to the retreat to be enjoyed at leisure. If the spider is not hungry, it may leave the 'parcel' in the web for a while.

Like all spiders, the orb spiders can absorb only liquid material, and they inject their victims with digestive juices to break down the tissues first. There is also a good deal of mastication by the bases of the fangs, reducing the flesh to a fluid and manageable consistency.

Nimble feet
The ability of the spider to speed around in its web without getting caught itself lies in the construction of its feet. Each foot ends in a hook-shaped claw, and when this is hooked over a thread the spider can glide down very easily, while a slight twist of the foot brings pressure to bear on the thread and holds the spider in any position. Special bristles around the claw guide the thread into the hook when the spider takes a new foothold, while an oily coating on the legs prevents them from sticking to the gummy threads of the trap.

Repair or renew?
The tough *Nephila* web may remain in a serviceable condition for many weeks, but the more delicate orb webs are generally renewed every day unless the damage caused by wind or prey is very slight or the spider has had a particularly good meal. It seems that minor damage does not trigger off the rebuilding instinct, and in some instances the damage may actually be 'repaired' – not by any conscious action, but simply as a result of the spider's dragging silken lifelines across the damaged part.

Complete renewal of the radii and spiral is the usual course, however, and this can be done remarkably quickly if the original bridge-thread and framework are still serviceable. Many spiders choose to carry out their rebuilding in the evening, but other species prefer the early morning – an indication perhaps of whether the species depends primarily on night-flying or day-flying insects.

Hammock and Sheet Webs

Although the orb web is certainly the most familiar kind of spider web, it is not the most numerous. That title must go to the more or less horizontal hammock webs produced by the members of the family Linyphiidae – the 'money spider' family. There are scores of these webs to the square metre in rough grassland and similar situations, and a single gorse bush may hold fifty or more. Most of these webs are only a centimetre or two across and they do not show up very clearly until covered with dew on an autumn morning, but then their shimmering beauty is displayed to its full effect and their vast numbers can be fully appreciated.

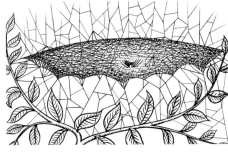

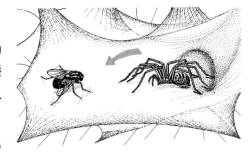

Deadly safety net
Left A typical hammock web with an aerial maze of trip-wires to snag unwary insects. Insects falling into the bowl are quickly despatched.

The domed hammock
Below Linyphia triangularis, spins a domed hammock web. Having no retreat, it clings under the web to paralyze falling prey.

Caught in the web
Above Prey caught in the hammock web of *Linyphia montana* is easy meat to the spider as it rushes onto the web from its retreat.

Death in the corner
Below A fly entangled on the sheet web of *Tegenaria* faces a swift death as the spider scurries from its funnel-shaped retreat.

How the hammock web works

The hammock web is a sheet ranging up to about 30cm (12in) in diameter and consisting of a rather haphazard arrangement of criss-crossing strands. Close examination shows that numerous threads rise from the upper surface to form what is best described as a maze of scaffolding attached to overhead twigs and leaves. The sheet itself may be slightly domed, it may be flat, or it may sag into the typical hammock shape.

The spider, typified by the abundant *Linyphia triangularis* – which must dwell in just about every bush and hedgerow in Europe, generally hangs upside down from the lower surface of the hammock, although some species prefer to rest in a nearby retreat. None of the threads is sticky as a rule, and the linyphiids depend on small flies and leaf hoppers that fall onto the sheet. It is here that the maze of scaffolding comes into play,

impeding the insects' flight and bringing them down within the spider's reach.

The criss-crossing threads of the sheet are like trip-wires to tiny insect feet, and the insects find it difficult to move off before the spider rushes along and plunges its fangs into them from below. The prey is then usually dragged down through the web and wrapped in silk before being eaten. The sheet is repaired later simply by trailing a further layer of threads all over the surface.

Cobwebs in the corner

The triangular cobwebs that develop so quickly in neglected corners are the work of house spiders of the genus *Tegenaria,* in the family Agelenidae. One corner of the sheet-like web is rolled up to form a funnel-shaped retreat, in which the spider spends most of its time. Except that there are few, if any scaffolding threads over the sheet, the web

works in just the same way as the hammock web already described.

Flies and other insects land on the web and get their feet entangled in the maze of trip-wires, and then the spider, sensing their struggles to get free, rushes out to give them several swift bites and carry them back to its retreat for dinner. *Tegenaria* does not wrap its victims in silk, and, unlike *Linyphia,* it runs on the upper surface of its sheet. It is still something of a mystery how it does this without getting entangled itself.

Species of *Tegenaria* live out of doors as well as in the house, and their webs can often be seen on old stone walls. The sheets are not usually very large in such situations because there is little overhead attachment for them, but the retreats often go deep into the crevices between the stones and the spiders can usually be seen only at night, when they emerge to sit at the tunnel entrances.

Sheet webs very similar to those of *Tegenaria* are common in rough grassland, especially around the bases of gorse and heather bushes. These webs belong to *Agelena labyrinthica* and they work exactly like the *Tegenaria* webs except that their efficiency is improved by a maze of scaffold threads above the sheet attached to surrounding foliage.

The evolution of the web

A study of these rather untidy sheet webs with their funnel-like retreats may provide some clues concerning the origin and evolution of webs. We can be certain that the earliest spiders were hunters, and it seems reasonable to assume that, like today's hunting spiders, most of them made silken retreats under stones

Halfway stage
Below The retreat of *Segestria* with its trip-wire pattern may represent a stage in the development of the beautiful orb webs.

Grassland sheet web
Right Agelena labyrinthica has a sheet web with scaffolding over it that acts as an obstacle course to trap flying insects.

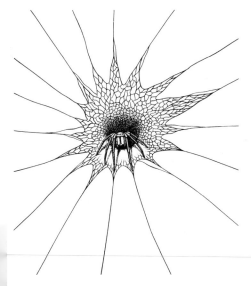

and in similar situations. Any tendency to leave strands of silk lying around the mouth of the retreat would have been an advantage to the spiders, for even if the silk did not catch wandering insects it would have alerted the spiders more quickly to their presence.

Thus we have a situation in which the 'untidiest' spiders — the ones leaving the most silk around their retreats — got the most food. Consequently there would have been considerable selection pressure in favour of more silk production, leading to the evolution of true webs around the retreats.

It is fairly easy to imagine the evolution of the sheet web from our hypothetical 'untidy' spider's retreat, and the hammock web can be derived from the sheet web without too much imagination. It is more difficult to imagine the evolutionary pathway of the beautiful orb web, but the possibility of such

evolution is there — perhaps via a stage resembling the simple but elegant web produced by *Segestria*.

The spiders have also explored several other design possibilities, some of which are shown on the next two pages. Web-spinning is clearly a very successful way of obtaining a meal, or not so many spiders would have indulged in it.

The spider's silk

Spider silk is a complex protein produced in glands in the animal's abdomen. There are actually several kinds of silk, each produced in a different type of gland. The glands occupy a large part of the abdomen, which is not surprising when one thinks how much silk is produced, and they are bathed with a rich supply of blood from which they extract the materials used to manufacture the silk.

The silk is secreted through the spinnerets, which project from the hind end of the body. There are three pairs in most spiders, although they are not always very obvious. Each spinneret carries many fine tubes, which are themselves connected to the various silk glands. The tubes are of different bores and they open onto the tips of the spinnerets in several different ways. The finest tubes and their openings are called spools and there are many of them to each spinneret. The coarser tubes project from the spinneret surface and are called spigots. There are rarely more than five of these on each spinneret and they are of several different shapes and sizes.

The complex connections between the various silk glands and spinnerets vary

The fine structure of a spider's spinnerets
Below This generalized spinneret shows the openings of the finer tubes — the spools — massed together in one area, with spigots of various sizes scattered all over the tip. Each kind of spigot extrudes a silk of a definite thickness and for a particular purpose.

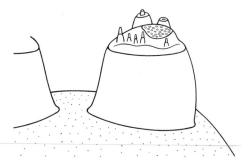

from species to species, but a few generalizations can be made as a result of scattered observations by a number of biologists. Silk exuded from the spools on the anterior spinnerets forms the short threads that make the attachment discs for the spider's lifelines, while the lifelines themselves and all the other long threads used in making webs emerge from various spigots. The sticky fluid that coats the web also comes out through the spigots, and so does the soft, yellowish silk with which the female wraps her eggs. This silk comes from special glands found only in the female. The broad bands of silk with which the spider wraps its prey emerge from the massed spools on the posterior spinnerets. All the different types of silk are liquid when they are produced, but they harden immediately on contact with the air to form the familiar threads that characterize spiders the world over.

Web-Throwing Spiders

The majority of web-spinning spiders simply make their webs—although, as we have already seen, the process is not really simple—and then sit back and wait for their meals to arrive. The abundance of web-spinning species in almost every habitat is testimony to the efficiency of the stationary web, but spiders are inventive little fellows and several species have devised ways of throwing their webs over their prey like the gladiators of ancient Rome.

Simple web-throwing

The daddy-long-legs spider (*Pholcus phalangioides*) indulges in a simple form of web-throwing. It spins a very flimsy and almost invisible scaffold web—in houses in southern England, but out of doors in many of the warmer parts of the world—and sits there motionless until dinner time. Other spiders form an appreciable part of the menu and, although they are not trapped by the flimsy web, their progress is impeded sufficiently for *Pholcus* to strike. Wisely, *Pholcus* does not get too close at first, but uses its long legs to throw strands of fresh silk over its victim from a distance. When the victim is sufficiently immobilized, *Pholcus* moves in and deals with it in the normal way. Mosquitoes and other small insects are dealt with in similar fashion.

This behaviour is, of course, only an extension of the normal spider behaviour of trussing its prey, but the various kinds of gladiator spiders really have gone in for something different, while the bolas and spitting spiders have developed the sports of angling and shooting.

Wrapping the prey
Below After throwing silk over it, *Pholcus* neatly binds up a fly with its back legs.

Preparing the net
Right Deinopis holds its partly completed net, made of sticky silken threads.

Gladiator spiders

Among the best known of the gladiator spiders are the various species of *Deinopis*. These are long-legged, brownish creatures that rival stick insects in their ability to conceal themselves so effec-

lively in the shady undergrowth.

Working on a rather flimsy, rectangular web, the spider produces a dense white net about the size of a standard postage stamp. The silk employed for the net is very elastic and very, very sticky. When the net is complete, the spider takes hold of a corner with each of its four front feet and settles down to wait. The hind legs take a firm grip on the framework of the larger web, and all eight legs are pulled in close to the body. When an insect comes within range, the spider immediately extends all its legs to their full length, thus stretching the net to several times its original size and at the same time shooting spider and net forward. With luck, the net will close around the prey and the spider will then despatch the victim with a quick bite behind the head.

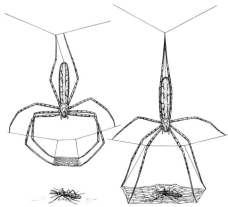

Deinopis spreading its net and snaring prey
Above left Suspended by two pairs of legs, the spider hangs motionless holding its elastic snare in its front four legs ready to catch prey passing beneath.

Above right As the prey passes underneath, the spider tensions the web-net, straightens its hind legs and drops the net down over its chosen quarry.

The spider is not always successful, however, and it may have to launch itself several times before making a capture. The supporting web sometimes breaks under the strain, leaving the spider hanging on its safety-line, and the spider may even drop its net, but it will carry on trying throughout the night, even if the net is severely damaged.

It seems that the spider will not make more than one net per night, but it will discard the net at daybreak even if it has not been used during the night. *Deinopis* usually leaps forward with its net, but some gladiator spiders release the net as they throw it over their prey.

Angling spiders

Bolas or angling spiders can be found in several parts of the world, but none is more striking than *Dicrostichus magnificus*, the so-called magnificent spider of Australia. This beautiful pink and cream creature takes up position on a twig at dusk and spins a short horizontal thread. Clinging to this thread, it then produces a second thread 5cm (2in) long and attaches one or more blobs of extremely sticky gum to the free end. This thread then hangs down from one of the spider's front legs, just like a fishing line hanging from a rod, and this is all the equipment the spider needs.

Sooner or later a moth arrives in the vicinity and, stimulated perhaps by vibrations of the moth's wings, the spider starts to whirl the weighted line round and round at high speed. If the sticky blob makes the slightest contact with the moth, the latter is doomed like a fly on a fly-paper. The spider hauls in the line and tucks into its meal, but there may be quite a battle before the prize is landed, with the moth straining to get away and the spider 'playing' it on the line just like a salmon fisherman. It is believed that the initial attraction of the moth to the spider's abode is due to a flower-like scent emitted by the spider.

A rather similar angling spider, known as *Cladomelea akermani*, lives in South Africa, but it holds its line with one of its hind legs and, instead of whirling it just when a moth appears, it whirls it continuously for about 15 minutes. It then hauls in the thread, bites off the globule and attaches a new one. Presumably, the globule begins to dry out and loses some of its stickiness as it hurtles through the air.

Spitting spiders

Amazing as the gladiators and anglers are, the prize for cunning and ingenuity must go to the spitting spiders of the genus *Scytodes*, including the little yellow and black *Scytodes thoracica* that lives in houses in Britain. These spiders creep about very slowly, usually at night, and seem incapable of over-powering anything of any size, but they have a secret weapon—a reservoir of powerful gum which can be fired out through the fangs. When the spider gets to within 6mm (0·25in) of a fly, it raises its head and fires: the victim is immediately pinned down by two zig-zag threads.

The zig-zag effect is produced by a very rapid oscillation of the fangs during the firing, and the threads hold the victim down so efficiently that *Scytodes* has no trouble in ambling over it and administering its deadly bite. When the prey is dead, *Scytodes* drags it away from the gum and consumes it at leisure. The gum, which is produced in glands in the much enlarged thoracic region of the body, is also used for defence against larger spiders.

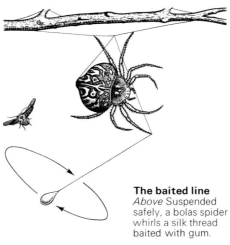

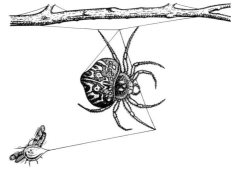

The baited line
Above Suspended safely, a bolas spider whirls a silk thread baited with gum.

A successful catch
Above A moth becomes stuck fast to the sticky blob and is hauled in to the spider's jaws.

Forming the bolas
Below Dicrostichus magnificus hangs a long silk thread coated with droplets of gum.

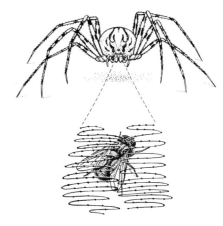

The ultimate weapon defeats a prey
Above The spitting spider, *Scytodes*, hunts small flies and other insects after dark. Creeping to within a few millimetres of its prey, the spider secretes two sticky lines from its chelicerae. These are thrown with a zig-zag motion over the intended victim and the gum secures it firmly to the ground.

Living Light Traps

Living light, or bioluminescence, is a feature of a large number of insects and also of various marine creatures such as squids (page 196) and certain deepsea fishes. The light is produced by a chemical reaction in which a material known as luciferin is combined with an enzyme known as luciferase. Oxygen is also bound into the combination, and energy is released in the form of light. When an ordinary electric lamp is switched on, only about 5 percent of the energy released is in the form of light. The rest is wasted as heat. In the luminous insects, however, as much as 92 percent of the energy is in the form of light – an amazingly efficient system producing light with virtually no heat. The light has several functions, but it is most often used for courtship and for the attraction of prey. Several of the angler fishes, for example, use luminous 'baits' to attract their prey (see page 58).

Femmes fatales

The most famous of the luminous insects are the glow-worms and fireflies, which are actually all beetles. Some fireflies use their lights for navigation, switching on as they come in to land and thus ensuring that they do not land on unsuitable ground. But the prime function of the light is in bringing the sexes together for mating. One or both sexes may give out light. The European glow-worm (*Lampyris noctiluca*) exhibits the phenomenon in its simplest form. The female is wingless and she sits quietly glowing among the vegetation waiting for a male to see the light and fly down to her.

Some of the fireflies behave in a similar way, but in many species the females are winged and both sexes give out light. They have developed some wonderful signalling methods to ensure that the right species meet up. The males flash their lights in a form of Morse code, with specific intervals between the flashes, and some trace distinctive patterns with their lights in the night sky. The females recognize the patterns of their own males and respond with their own lights. The females of many American species of *Photinus* reply to the males' signals after a very definite interval of time – usually between one and five seconds.

The fireflies rarely eat much in the adult state, but the females of *Photuris* are notable exceptions to this rule. They carry on the carnivorous habits of the larvae and they sometimes invade spider webs to eat glowing fireflies that have already been wrapped by the spiders. More fascinating is the way in which *Photuris* females have developed their courtship signals into effective ways of catching food. Before they have mated,

the females respond to male *Photuris* flashes in the normal way, but after they have mated they start to reply to the flashes of various species of *Photinus* fireflies. The males come down to investigate and they are caught in the jaws of the larger *Photuris*.

Professor James Lloyd has studied *Photuris versicolor* in Florida and has discovered that a single female can respond correctly to at least three species of *Photinus* by adjusting the time delay between their flashes and hers. In effect, the *Photuris* female can 'read' three different languages.

Cave spectacles

The Waitomo Caves of New Zealand have been turned into world-famous attractions through the activities of a small maggot known as the New Zealand glow-worm (*Arachnocampa luminosa*). Myriads of grubs, each only a centimetre or so in length, live on the ceilings of the caves and illuminate them each night with an eerie, greenish light. Similarly illuminated caves are found in various parts of Australia.

The insects do not, of course, give out their light to attract human visitors. They are interested in much smaller visitors – the swarms of midges and other insects that live in the caves. The glow-worms, which themselves are the larvae of small flies, construct silken snares that hang from the ceiling like bead curtains. Each bead is a blob of sticky mucus. Each larva clings to a horizontal thread on the ceiling and gives out its cold light. The light is reflected by the millions of tiny mucus globules on the threads, giving the whole roof a shimmering appearance. Small flies are attracted to the lights, just as they are to our electric lamps, and many become entangled in the snares. The glow-worms then pull up those threads that have been successful and consume the victims.

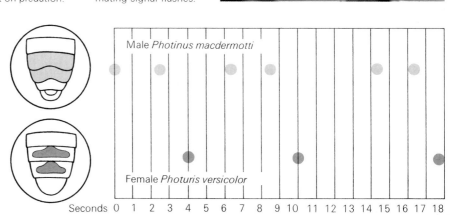

Invitation to death
Below The double courtship flashes of a male firefly are answered by a 'false' reply from the female of a species bent on predation.

Successful deceit
Right A female firefly *Photuris versicolor* has captured the smaller male of *Photinus tanytoxus* by replying to his mating signal flashes.

Male *Photinus macdermotti*

Female *Photuris versicolor*

Seconds 0 1 2 3 4 5 6 7 8 9 10 11 12 13 14 15 16 17 18

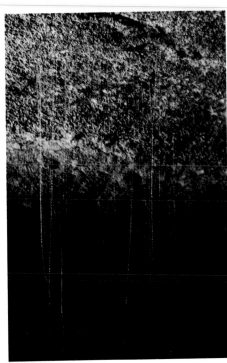

Predatory chandelier
Above The roof of the Waitomo Caves in New Zealand offers a breath-taking spectacle to the visitor. The thousands of tiny lights are the luminous lures of predatory glow-worms *Arachnocampa luminosa.*

Shimmering snares
Right A closer view of the cave roof reveals clusters of hanging threads festooned with sticky globules. These are produced by the glow-worms to snare inquisitive insects.

A silken hammock
Left The glow-worms are about 1cm (0.4in) long and are suspended close to the cave ceiling in hammocks made from silk. They lie still until prey becomes trapped in their snares, when they haul in and consume their victims.

Anglers and Archers

In this final look at animal 'trappers' we will consider a few animals that do not actually make traps or snares but which nevertheless employ special methods to bring prey to them. They show certain parallels with the human fisherman and the huntsman with his gun or bow and arrow in search of likely victims.

The angler fishes

It does not take much imagination to see how the angler fishes got their name; almost all of the 350 or so species possess a fishing rod complete with bait dangling in front of the mouth, and they feed almost entirely on other fishes. Angler fishes are often regarded as purely deep-sea creatures, but only about a third of the species live in really deep water. Most of the others live in the relatively shallow waters over the continental shelves. As a group, they are really rather ugly fishes, seemingly all head and mouth. The largest is little more than 1m (39in) long, and yet its head and mouth are about 75cm (30in) wide — a truly grotesque creature but, as we shall see, one with an efficient, functional design for its chosen habitat.

Angler fishes spend nearly all of their time sitting on the seabed expending very little energy, so it is not surprising that their gills are very small. They could not chase their prey if they tried, for they are very poor swimmers with pectoral and pelvic fins that look more like hands and feet than efficient paddles.

The fishing rod is actually one of the spines of the dorsal fin that bends forward over the head. It carries the bait on its free end. The bait itself is a flap of skin, but its shape varies according to the preferred prey of the angler. Most anglers have a red, worm-like bait that is very attractive to small fishes. Anglers living in the blackness of the depths often have luminous baits to attract their prey. Some anglers keep the rod permanently at the ready, but others can fold the rod away in a groove in the head when they are not hungry and not actively 'fishing'.

When fishing, the angler waves the bait gently to and fro through the water until it has attracted the attention of a small fish. As the fish comes closer to investigate, the bait is lowered gradually towards the angler's mouth, and the prey usually follows it. Unlike the human angler, however, the angler fish does not want the bait to be taken, and at the last moment it whips the bait away from the prey. At the same time the great mouth opens and the inward rush of water carries the victim to its doom. The jaws are snapped shut again in an instant — far too quickly for the human eye to follow the movement — and a battery of backward-pointing teeth prevent any possible escape for the prey.

Quite large prey may be caught when the opportunity arises; one large angler fish (*Lophius piscatorius*) has been caught with its stomach filled with a cod measuring 60cm (24in) in length. It seems likely that both sight and mechanical stimulation of the fishing rod are involved in the capture of prey, but the angler is not a very discriminating predator and it will swallow virtually anything that comes along.

Although the bait or lure of the angler fish is normally very conspicuous, the rest of the body is certainly not. It is frequently clothed with an assortment of frilly lobes and tassels of skin that make it very difficult to pick out as it rests among the seaweed. Some species can even change their colours to match different backgrounds.

One group of anglers known as bat-fishes take the camouflage even further. Their skin carries so many filamentous outgrowths that the fishes look just like clumps of seaweed. Other fishes are fooled by this and they come up to nibble the anglers. The fishing rod, which has been hidden until this time, now comes out and the bait lures the inquisitive visitor away from the angler's skin and towards the mouth. The rod may be alternately protruded and withdrawn several times, 'playing' the victim like a salmon fisherman, but eventually the prey is drawn near enough for the angler to strike with its jaws.

The hunting angler

The sargassum weed, a seaweed that

An angler 'plays' a fish to its death

Below The angler fish attracts a curious fish by gently waving the lure above its head.

Both fish and lure are gradually brought closer to the wide jaws of the camouflaged angler.

Suddenly the huge, broad mouth is opened and the inrush of water sweeps the fish to its death.

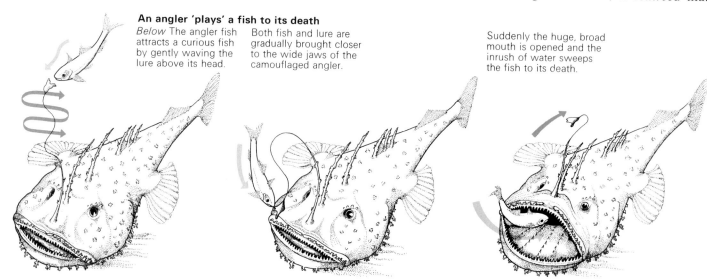

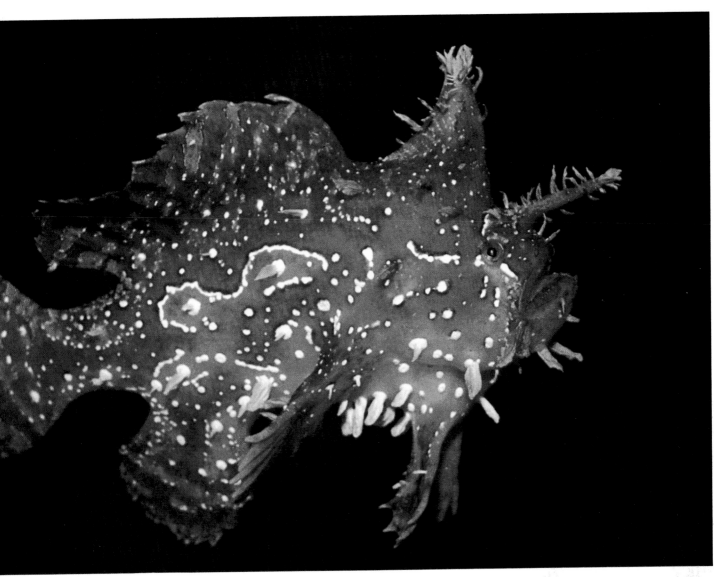

drifts in vast masses far out into the western Atlantic, is home to one of the strangest of all the angler fishes. This is *Histrio histrio*, in which the female reaches lengths of about 15cm (6in) while the male is a mere 2·5cm (1in) long. *Histrio* has given up angling, although it is clearly related to the other angler fishes, and it hunts shrimps by stalking them slowly through the clumps

of weed. Its pectoral and pelvic fins are amazingly hand-like, even to the extent of having 'fingers', and the fish crawls along the seaweed strands in very much the manner of a chameleon crawling along a branch. Its camouflage is truly remarkable and it is almost impossible to pick out the hunter in the midst of a clump of weed. The shrimps certainly don't see it coming, for it creeps right up

to them until it can gulp them down into its typically large mouth. *Histrio* has clearly mastered its strange environment, for there are few clumps of sargassum without at least one of these fascinating angler fishes.

Fishing turtles

The angler fishes are not the only creatures to use baits to lure other fishes to their deaths, although no other creature has such a well-developed rod. The alligator snapper turtle (*Macrochelys temmincki*) lives at the bottom of muddy rivers in the eastern United States and, almost buried in the mud, it lures small fishes with the aid of its strange tongue. This is forked to form two pink, worm-like branches which show up very clearly against the dark background of the open mouth. This lure is moved about and proves irresistible to small fishes, which are eagerly snapped up and swallowed with little chewing.

The South American matamata (*Chelys fimbricata*) employs similar methods, but here the bait is a series of fleshy filaments around the edges of the mouth and along the sides of the neck.

Camouflaged creeper
Top The grotesquely camouflaged angler fish, *Histrio histrio*, blends into the sargassum weed amongst which it lives, using its finger-like fins to crawl slowly in search of shrimps. It has no lure but retains the large mouth found in all angler fishes.

A tempting tongue
Right The alligator snapping turtle lies motionless in the water with its mouth open to reveal a pink tongue that resembles a tasty worm to a hungry fish. Any that are unwary enough to take the bait are swiftly swallowed.

Spiders that fish

Spiders of the genus *Dolomedes* are closely related to the wolf spiders (see page 82) and many of them hunt in the same way, although they prefer to live in marshy areas and even on slow-moving water. They are frequently called raft spiders, because it was once thought that they made rafts of leaves and floated about on them. With more justification they could be called fishing spiders, for several of the larger species, such as *D. plantarius* and *D. triton*, do catch small fishes on quite a regular basis.

These spiders live in swamps and pools and spend much of their time sitting on the floating leaves of pondweeds and other aquatic plants. Close examination reveals that the front legs of the spiders are generally resting on the water surface, where they can pick up vibrations of anything moving nearby. Insects struggling on the surface can be detected by this means, and so can other animals swimming in the water below. A spider may streak down into the water and re-emerge with a fish that it has detected and caught.

This is, of course, nothing more than plain hunting, but the raft spiders often take things a stage further and attract prey towards them. They do this by sitting on floating leaves in the normal way but, instead of merely resting their legs on the water surface, they dabble them gently into the water. The movements attract small fishes as surely as the angler's worm, and the spider can then pounce with very little effort. Like the other wolf spiders, the raft spiders have good eyes and they use them in at least the later stages of prey capture.

The fish that spits bullets

The mangrove swamps that fringe the coasts of Southeast Asia and northern Australia are home to the amazing little archer fishes (*Toxotes jaculatrix* and its relatives). These fishes catch much of their food in the water just like other fishes, but when they are hungry they literally shoot down insects on overhanging vegetation with a hail of water droplets fired from the mouth just like bullets.

There is a deep groove in the roof of the fish's mouth, and when the tongue is pushed up to meet it a narrow tube is formed. This tube is about 2mm (0·8in) in diameter and it functions as the blowpipe through which the water is fired. The eyes of the archer fish are very efficient and, being forward-looking, they give a good field of binocular vision (see page 68). Observations suggest that the archer fish can pick out a small insect at least 2m (6·5ft) away, and the binocular vision allows it to judge the distance

Walking on water to catch prey
Above The raft spider senses the ripples from a mayfly struggling in the surface tension.

Sensing the intervals between ripples allows the spider to pinpoint the prey's position.

Anchored to the floating leaf by a silk life-line, the spider pounces on the defenceless victim.

Dabbling for a bite
Above Sometimes the raft spider will dabble its front legs into the water to attract fish.

As a fish comes to the surface to investigate, it is swiftly grabbed and bitten by the spider.

The bite injects a paralyzing venom and the rigid fish is easily hauled out and eaten.

to the intended prey very accurately.

Having picked out a possible victim, the fish manoeuvres itself into a clear firing position, rising up until its mouth just breaks the surface. The tongue is raised to form the firing tube, and a sudden compression of the gill chambers forces water into the mouth and out through the tube. The tip of the tongue

acts as a valve, allowing the fish to fire single bullets, a stream of separate bullets, or a continuous jet of water.

An adult archer fish is a very good shot over distance up to 2m, and the insect soon falls into the water under the constant bombardment. It is then snapped up by the fish. As well as being very accurate, the fish has considerable

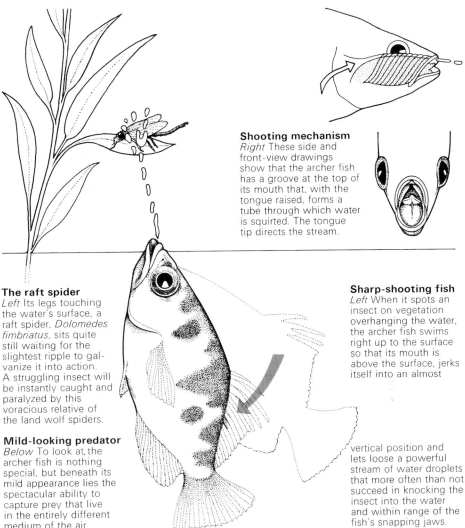

Shooting mechanism
Right These side and front-view drawings show that the archer fish has a groove at the top of its mouth that, with the tongue raised, forms a tube through which water is squirted. The tongue tip directs the stream.

The raft spider
Left Its legs touching the water's surface, a raft spider, *Dolomedes fimbriatus*, sits quite still waiting for the slightest ripple to galvanize it into action. A struggling insect will be instantly caught and paralyzed by this voracious relative of the land wolf spiders.

Mild-looking predator
Below To look at, the archer fish is nothing special, but beneath its mild appearance lies the spectacular ability to capture prey that live in the entirely different medium of the air.

Sharp-shooting fish
Left When it spots an insect on vegetation overhanging the water, the archer fish swims right up to the surface so that its mouth is above the surface, jerks itself into an almost vertical position and lets loose a powerful stream of water droplets that more often than not succeed in knocking the insect into the water and within range of the fish's snapping jaws.

power; shots which have missed their target, as a certain proportion are bound to do, have been known to travel at least 5m (16·5ft), and inquisitive observers have found that the stream of bullets produces a sharp pain if it hits the face.

Archer fishes learn to shoot when very young, although the youngsters are not very accurate and their range is limited to a few centimetres. Range increases quite naturally as the fish gets larger and stronger, but accuracy has to be achieved by practice or learning.

The archer fish is thus thought to be one of the more intelligent fishes, but one thing that puzzled biologists for a long time was its apparent ability to overcome the problem of refraction — the bending of light rays at the air/water interface. If a pencil is put into a glass of water it will appear to bend because of the bending of the light rays, and when viewed from above the bottom of the pencil will appear to be nearer than it really is. If viewed from under water, the top of the pencil will appear to be further away than it really is, and the same must apply to the insects seen by the archer fish.

It used to be thought that the fish had an inbuilt compensatory mechanism

that allowed for this when taking aim, but careful observations eventually showed that the fish actually avoids the problem instead of solving it. When it spots a possible target, the archer fish swims until it is very nearly underneath it. This reduces the refraction to negligible proportions because light passing between air and water perpen-

dicularly to the interface is not bent at all. The fish can thus aim directly at the target.

Just before firing, the fish jerks its body into a near-upright position so that its big, forward-looking eyes and mouth are more or less in line with the target. This allows it to judge the distance accurately before firing.

Meat-Eating Plants

We have seen in the introductory section that green plants are the primary producers of food on this planet and that all animal life depends on them. Plants are basically passive organisms and, although some, such as the stinging nettle, have evolved protective devices, it is the normal order of things for animals to eat the plants. It might, therefore, come as a surprise to learn that some plants have turned the tables on the animal kingdom and actually catch and 'eat' animals.

Man-eating plants belong to the realms of science fiction, but insect-eating plants are real enough and they possess some really amazing adaptations that help them to catch and digest their victims. Some, such as the butterwort and the aquatic bladderwort, rely on small creatures blundering into them, but the majority lure their victims to their death with bright colours or scents and their traps are every bit as efficient as those of the animals we have just seen.

Enriching a restricted diet

Most insectivorous plants grow in mineral-deficient soils and a large proportion of them grow in peat bogs. The acidic conditions of a peat bog slow down the rate of bacterial decay, thus allowing the peat to build up from the dead plant remains. Nitrates and other essential minerals are locked up in the peat and very small amounts are available for growing plants. Any feature which helps the plants to overcome this problem is clearly of great value, and meat-eating certainly comes into this particular category.

Plants of several different families have evolved insect-catching habits which help them to thrive in the mineral-deficient soils, but they do not depend entirely on insects. The plants contain chlorophyll and carry out photosynthesis in the normal way, and many can grow without any insect food at all, especially when grown in a mineral-rich soil. They tend to be very stunted on poor soils, however, and it is clear that the insects provide them with that vital extra mineral supply to enable them to grow well. Deprived of insect food, many of the insectivorous species fail to flower.

A vegetable gin trap

Probably the best known of the insectivorous plants is the Venus' fly-trap (*Dionaea muscipula*). This is a native of bogs and other damp, mossy places in North America, but it is widely grown as a fascinating house plant. The leaves are arranged in a rosette and the leaf stalks are broad and flat and they carry out photosynthesis in the usual way. The blades, however, form the traps.

Each blade is more or less circular and distinctly hinged along the midrib. The margins bear long teeth, and the green outer zone of the blade carries insect-attracting nectaries. The inner part of the leaf is covered with red digestive glands, whose colour also helps to attract insects, and each half of the blade also bears three prominent trigger hairs. These are the 'brains' behind the operation and they can actually count. A single stimulation of one of the trigger hairs, such as it might receive from a falling rain-drop, produces no reaction, but a second touch following in quick succession springs the trap: the two halves of the blade snap together within a second and the marginal teeth interlock. Simultaneous stimulation of two of the trigger hairs will also cause closure of the trap.

In the wild, the closure stimulations are most frequently provided by an insect that has been attracted to the trap by the colour or nectar, and the insect is then immediately imprisoned by the teeth. It is possible to fool the plant by tickling the triggers with a piece of grass, but you will not fool it for long because it needs the chemical stimulus of an insect body or some other meat to keep the trap shut.

If it does catch an insect, the closure does not stop when the teeth have met: the two halves become pressed very tightly together and the red digestive glands begin to pour out their juices. These contain protein-splitting enzymes which break down the bulk of the victim's body and dissolve it so that the plant leaves can absorb it.

Charles Darwin — always thirsting for knowledge of nature's miracles — cut a small hole in the bottom of a closed trap and discovered that digestive juices continued to flow for nine days. When digestion has been completed, the leaf blade opens again and the remains of the victim, now no more than an empty shell, blow away in the breeze.

The sticky sundews

The sundews, which belong to the same family as the Venus' fly-trap, are found in bogs and on wet heathlands nearly all over the world. They are mostly small plants with rosettes of ground-hugging leaves. The leaf blades may be round, spoon-shaped, or long and narrow, but they are all clothed on the upper surface with reddish tentacles. Each tentacle has a swollen tip which secretes a shiny and very sticky fluid, making the plant look as if it is covered with dew. Small insects are attracted by this dazzling array, and probably also by the mild fungus-like smell of the secretion, and become firmly stuck to the gluey tentacles.

Even the lightest insect sets up pressures and stresses on the tentacles, and this causes them to bend towards the centre of the leaf. Neighbouring tentacles also bend over, although it is not known how they are stimulated, and the insect is pressed against the shorter tentacles in the centre of the leaf. The tentacle heads then secrete protein-splitting enzymes, which dissolve the body of the victim. The sundew leaf absorbs the proceeds.

If a large insect is caught and it struggles violently, all the tentacles may bend over it and the whole leaf blade may curl up to enclose it. The leaf and tentacles straighten out again when digestion of the victim is complete and the soluble food products absorbed. They do not begin to secrete the sticky fluid again until they have dried off and the horny remains of the previous meal have blown away.

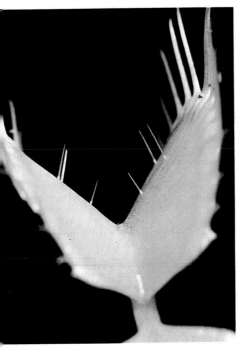

The Venus' fly-trap
Above and left The
Venus' fly-trap has
hinged leaves modified
to form a spiked
gin trap. Situated
in the centre of each
half of the trap
there are small trigger
hairs that when
disturbed by a visiting
insect alter pressure
within the hinge cells
and cause the trap to
snap shut. The long
teeth interlock
preventing any escape.

The sticky sundew
Right The sundew has
a rosette of stalked,
lobed leaves covered
in sticky tentacles.
Insects attracted to
the sugary solutions
secreted by the plant
are stuck down as soon
as they alight on the
leaves. The lobes then
curl up around the
unlucky victim.

Pitchers of death

The pitcher plants use bright colours and nectar to lure flies and other insects to their deaths in pools of digestive juices. These fluids are contained in modified leaves called pitchers, some of which are shaped like ice-cream cones and some of which look more like old-fashioned jugs. Both kinds generally have some kind of lid or canopy to keep out the rain.

The cone-shaped pitchers belong to a genus of North American plants known as *Sarracenia* that live in boggy and marshy areas. The pitchers come straight up from the ground and are often brightly coloured, although each one generally has a green 'wing' on one side where photosynthesis takes place. The rim of the pitcher is turned in to form an inward-pointing lip, on which there are numerous honey glands. Insects that stay on the lip are safe, but most visitors enter the funnel in search of more food. And then they are doomed.

Just below the lip there is a very slippery region whose surface is covered with loose, waxy scales: the insects cannot get a grip and they slide down through a region of downward-pointing hairs and into the liquid, which fills about one third of the pitcher. Unable to climb back through the downward-pointing hairs, the insects remain in the liquid, which in some plants is thought to contain a mild anaesthetic, and they soon drown. Digestive juices quickly break down their dead bodies,

and the plant absorbs the useful minerals.

Nepenthes is a genus of pitcher plants from the tropics of Asia and Australia. Some of the plants are rooted in the swampy soil, with long climbing stems rising up through the trees, but many grow perched on the tree trunks and branches as epiphytes. They do not take food from the trees and, although the droppings of birds and other animals provide some minerals, this is another situation in which minerals are in short supply and in which the ability to catch insects is a great advantage.

Nepenthes is distantly related to *Sarracenia*, but it forms its pitchers in a very different way. The midrib of the leaf is prolonged into a slender tendril, which then expands to form the jug-shaped pitcher. This can be quite large

and hold well over a litre of fluid in some species, although the pitchers are rarely more than about one third full.

As in *Sarracenia*, the rim is turned in to form a lip on which there are numerous nectaries. There are also nectaries on the underside of the lid, and there is no shortage of insect visitors. Once the insects enter the pitcher and reach the waxy surface they slide down into the acidic liquid and are digested. The digestive juices are so strong that they can even dissolve egg shells as well as insect cuticles.

Not all the insects lured to the pitcher end up in the fluid, however, for a number of small crab spiders have discovered an easy way to make a living. They anchor themselves just inside the pitchers and snap up a proportion of the insects that

Leaves as pitchers
Right Pitcher plants, such as this type of *Sarracenia*, have each leaf modified into a bowl-like trap with a protective lid. Insects are lured into the bowl in search of the sugar solution secreted by special glands called nectaries. As they crawl deeper into the bowl, the insects suddenly slip on a covering of loose waxy scales and tumble into the pool of acidic digestive juices filling the bottom third of the pitcher. This converts the prey's tissues into soluble food.

Tropical Nepenthes
Left The lip of this pitcher shelters the bowl and its digestive juices from tropical rainstorms. Insects attracted to the nectaries are prevented from escaping by the slippery surfaces and incurved spikes near the rim.

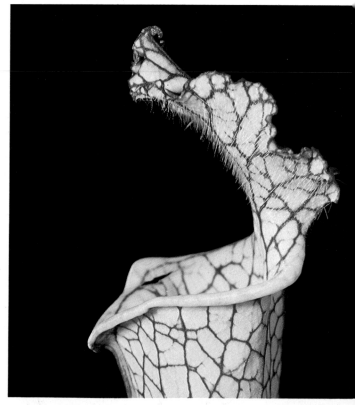

Trapper thief
Right This crab spider makes its home in the rim of the pitcher trap. Here it can live in the comfortable, protected environment of the pitcher plant and feed on insects attracted to the sugary bait secreted by the pitcher's nectaries. Being an aggressive hunter, the spider will not tolerate another spider in its territory, keeping all the benefits to itself. The raiding of its food resources is not harmful to the pitcher plant as it has several traps each attracting many insects. The spider is immune to the digestive juices in the base of the trap.

drop in. The spiders are to some extent immune to the digestive fluids and when disturbed — by an inquisitive biologist, for example — they drop down and submerge themselves for a few minutes.

The ticklish bladderwort
Most species of bladderwort live in water, where they float freely just below the surface without any kind of root to anchor them. The bladders from which the plants get their name are scattered along the hair-like stems and leaves. They are like little flat purses, only a few millimetres long and with an opening at one end. This opening is in the form of a tiny trapdoor surrounded by a number of very fine, branched trigger hairs. The pressure inside the bladder is somewhat reduced, and when

a water flea or some other small creature tickles the trigger hairs the trapdoor flips inwards and the rush of water into the bladder sweeps the luckless victim with it. The door then closes and the captured animal soon dies, although it is doubtful if any digestive juices are produced inside the bladder.

The victim's body gradually decays, and the products of decay are absorbed by cells lining the bladder. Some water is absorbed as well, thus reducing the internal pressure and resetting the trap in readiness for the next victim.

Murderous fungi
The plants featured elsewhere in this section are all flowering plants, but flowering plants do not have a monopoly when it comes to killing animals. A tiny,

hair-like soil fungus called *Dactylella* branches out to form minute rings that catch eelworms. As soon as an eelworm passes through one of the rings, the ring contracts and holds the animal firmly. New fungal branches then grow into the worm's body and digest it.

Related soil-dwelling fungi catch eelworms and other tiny soil animals by producing very sticky branches. Large insects can also be killed by fungi, although it cannot be said that the fungi catch the animals. The reverse is nearer the truth, for the fungal spores get into the mouth or breathing pores of the insects and start to grow. Eventually, the insect's body becomes filled with solid masses of fungal threads and new spore-bearing branches grow out, giving the corpse a hairy appearance.

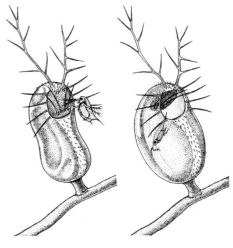

Bladderwort — the aquatic suction trap
Above When a small animal blunders into the trigger hairs on the lid of this trap, the door springs inwards and breaks the vacuum inside. The walls, which were held together, spring apart and the prey is sucked in.

The sticky carnivore
Left The butterworts of the northern hemisphere and South America are found on boggy and marshy ground. They secrete a yellow sticky substance on the leaves and stem resembling butter, hence the name butterwort. Insects are attracted both to the purple flower and to attractive glistening secretions on the leaves and stem. When a fly lands on the sticky leaf it is immediately glued down. Its struggles trigger gland hairs on the leaf's surface to secrete digestive enzymes that are acidic and dissolve the tissues of the trapped fly. The soluble food is then absorbed into the tissues of the plant. Being carnivorous, the plants can survive in poor soil of little nutritive value.

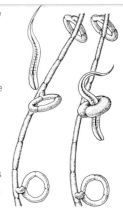

The living snare
Above The soil fungus *Dactylella* has minute snares at intervals along its threads. When an eelworm wriggles through one of these rings the sections swell up reducing the size of the hole and closing tightly around the prey. Fungal branches then invade and digest.

Strategy and Weapons

Predators which find their prey by means of contact receptors do not really chase their prey, but their hunting methods are not always as random as they might appear. Young ladybirds, for example, follow a very definite pattern as they search plants for aphids, and this pattern ensures that they cover virtually every part of a plant.

The strategy of the chase

Arriving on a fresh plant, the ladybird larva moves upwards and, if it does not find food, it continues upwards until it can go no further. If food is still not found at the top of the plant — the most likely spot for a cluster of juicy aphids — the insect moves down a short way and then goes up again. It may, of course, go up the same shoot again, but the chances are that it will crawl up a side shoot or leaf. If the ladybird is still out of luck it will climb down even further and then ascend again. This process goes on until, if it finds no food, the ladybird reaches ground level. By this time it will have climbed up most, if not all, of the shoots and branches and scoured virtually the whole plant. It will then go off to find another plant.

Sooner or later the ladybird will find a juicy cluster of aphids, explore them briefly with antennae and mouthparts to ensure palatability (a few aphids are poisonous to ladybirds), and then set to work with its jaws. These are strong and pointed and they easily pierce the skin of an aphid. The younger ladybird larvae inject saliva into their prey and then suck up the partially digested contents, but older larvae and adults simply chew up the complete aphids.

When it has found and eaten an aphid, the ladybird does not walk straight on: it changes direction after every few steps and this behaviour keeps it in a small area — just what is required if it is to make the best use of a cluster of aphids. When the supply is exhausted, the ladybird reverts to its up and down hunting procedure.

These behavioural patterns are entirely instinctive, but they compensate admirably for the insect's inability to detect its prey from a distance and ensure that it does find food.

For animals that see, hear, or smell their prey from a distance, the detection of the prey is only the beginning of what could be a long chain of events. Signals from the sense organs reach the brain, which then decides — perhaps purely by instinct or perhaps partly as a result of past experiences — on the correct course of action. Instructions are then sent out to the muscles to carry out the appropriate actions for the chase to begin. Each predatory species has its own

The aphid's enemy
Above This larval 7-spot ladybird feeds voraciously on aphids. It eats almost continuously until it pupates and it makes a systematic search of the vegetation (see left) in its quest for prey. It climbs as high as it can to start with, unless it finds food straight away, and then comes down before going up a new branch.

hunting strategy. There may be a straightforward, high-speed chase, such as we find among the hawks and other birds of prey and, in fact, all the aerial hunters, but more often the prey is stalked until the predator is within striking distance. This is the method used by geckos and chameleons, snakes, jumping spiders, and most cats. Some predators, however, use a combination of methods, stalking up on their prey to start with and then breaking into a

high-speed run. The cheetah — the fastest animal on four legs — is the best exponent of this kind of hunting, but the same principle is used by several of the social mammals, such as the wolf, the lion, and the hunting dog.

Social hunting, in which several members of a species combine to chase and corner prey, is found mainly in mammals, but examples can also be found among the pelicans (page 184), some ants (page 122), and the sharks (page 186), where the thresher sharks round up smaller fishes much as cowboys round up cattle.

Weapons for the kill

Teeth, jaws, and claws are the main weapons used for catching and killing prey, but it is not always possible to separate the action of catching the prey from the act of killing it. Many of the more powerful predators use their formidable weaponry to catch and kill their prey simultaneously. A pigeon plucked from the air by a peregrine's talons, for example, is not just caught, but killed as surely as if shot with a gun. Similarly, the final spring of a leopard or a tiger brings down its prey and very often breaks its neck in the process.

If the prey is not killed by the first blow, it is very soon despatched by the deadly claws of the hunter or by the teeth that close around its neck as it falls. Large, stabbing 'canine' teeth are characteristic of all the land-dwelling carnivorous mammals and they play a major role in killing the prey and ripping its flesh.

One can, in fact, learn a great deal about the diets of mammals simply by examining their teeth. Grazing mammals, for example, have flat, grinding surfaces to their teeth, while fish-eating mammals such as dolphins have simple peg-like teeth designed merely to hold the slippery prey. In cats and dogs especially, sharp-edged cheek teeth, known as carnassials, slice the flesh into smaller pieces before it is swallowed.

Birds of prey use their powerful talons to catch their prey, and employ their hooked beaks only to tear the flesh afterwards. Most other birds, however, use their beaks to catch their food and the shape of the beak varies according to the diet. Snakes and fishes normally use their teeth to capture their prey, although other parts of the body may be used as well, as with the thresher shark and the sawfish. Chameleons and some frogs and toads trap their prey with long, sticky tongues. Many predatory beetles possess powerful jaws armed with strong teeth, although the teeth are of a very different nature from those of vertebrate animals, while the predatory bugs have sharp, piercing beaks which they plunge into their victims to hold them securely while sucking out their juices. Dragonflies use their spiky legs like nets to scoop smaller insects from the air.

Many predators overcome their prey through their superior size and strength. This applies to such diverse creatures as tigers, eagles, kingfishers, sharks, lizards, many snakes, and predatory beetles. Biting the prey, or merely holding it in the jaws is usually sufficient to subdue the prey if not to kill it before it is eaten, but the constricting snakes, which frequently take prey much bulkier than themselves (see page 94), bring the whole body into play to subdue the struggling victim.

A number of predators employ poison to kill or paralyze their victims, the best known poisoners being various snakes. Some of the salivary glands are modified to produce venom, and some of the teeth are modified to a greater or lesser extent

Tools of the trade
Hunters come in all shapes and sizes, each having its own particular method of despatchings its prey. The shape of the structures used in killing the quarry reflects the hunter's habits and environment. *Right* The assassin bug sticks its long beak into its prey to inject a dose of paralyzing and digestive saliva. *Below* Lions despatch prey by sinking their long teeth into the neck and suffocating it. *Bottom* Owls have powerful grasping talons to catch prey.

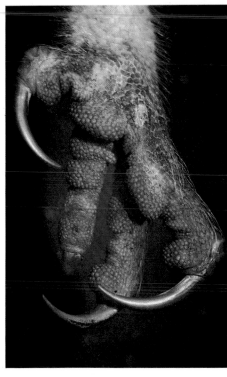

to inject it into the victim when the snake bites (see page 94). Spiders and centipedes also possess venomous glands and fangs to immobilize their prey, while the wasps have poison stings.

Perhaps the most surprising weapon used by the hunters is electricity. A number of fishes surround themselves with electric fields or give out streams of electric pulses as an aid to navigation in murky waters (see page 190), and several actually find their prey by following up disturbances of the electric field. This mechanism has evolved to its logical conclusion in the electric ray (*Torpedo*), where powerful electric discharges are produced to stun the prey species so that they can easily be picked up and eaten.

The remaining pages of this book are devoted to the detailed methods by which a wide range of vertebrate and invertebrate hunters capture their prey on land, in the air, and in the water, with a section on group hunters.

Hunters on the Land

Scent, sight and sound lead terrestrial hunters to their chosen prey — speed and stealth allow the predators to reach their victims — strength or speed, sometimes assisted by venom, effects the kill.

In this section of the book we will be taking a look at a selection of land-based predators. Later sections will deal with predators in the air and in the water and look at the specializations required for detecting and hunting prey in these environments.

The land-based hunters include a very wide variety of predators, from small insects and spiders, through strange slugs and snails, to the top predators such as the lion and the tiger. Between them, they employ all the major methods of tracking down, catching, and killing their prey. Chemical detection is especially important among many of the smaller predators, where eyes and ears may be very poorly developed or even completely absent. Scent is also important for many hunting mammals and some snakes, but most of these larger animals use their eyes and ears as well. Some snakes detect the warmth radiating from their prey.

Speed and stealth

The two main features involved in approaching prey on land are speed and stealth. The high speed approach is exemplified by the cheetah (page 110) and, among the lower echelons of the hunters, by the wolf spiders. Wolves and hunting dogs also capture their prey through speed, although here it is stamina and social cooperation rather than absolute speed that combine to defeat the prey after a chase. Those predators that cannot outrun their prey must employ the stealthy approach.

Stealth necessarily involves a certain amount of camouflage on the part of the predator whose prey has good eyesight, and this is beautifully shown by the tiger. The vertical stripes on the body break up its outline and enable it to merge very effectively with the tall grasses through which it often stalks its prey. The spotted coat of the leopard similarly breaks up its outline and disguises it among the dappled light of the forest, while some beautiful examples of camouflage can be found among the snakes and jumping spiders. Even the plain-coloured lion is quite well camouflaged in its natural habitat, especially when one realizes that its prey is colour-blind and sees everything in shades of grey — again, we must relate the abilities of a predator to those of its prey and not to our own abilities if we are to appreciate its behaviour to the full.

But camouflage is not confined to the hunter; the hunted have evolved their own forms of camouflage in the never-ending battle for survival and even large antelopes out in the open can be difficult to see from a distance. Some have disruptive stripes that break up the body outline, but most employ a system known as counter-shading. The underside of the body is lighter in colour than the back, and this helps to even out the shadows and 'flatten' the animal, thus merging it with the background.

If they had to rely on sight alone, many of the large predators would go hungry, but, of course, they also use their ears and noses to help guide them to their prey. The prey animals also have ears and noses, however, and the predators must allow for this if they are to be successful. Most of them can approach in virtual silence, but their own scents can be a problem and many a big cat has lost its dinner through approaching upwind of its prey and allowing its scent to precede it and warn the prey. It seems difficult for animals that hunt primarily by sight to learn to approach from downwind, although many certainly do circle round when necessary and come in from the right direction. Animals that hunt primarily by scent automatically approach from downwind, because they pick up the scent of their prey in the air and turn to face it. This enables predators to creep close to potential prey before pouncing.

The catch

Many predators catch and kill their victims by sheer strength, while others use venoms to kill or at least paralyze their prey. In both groups there is usually a well developed technique for administering the final coup-de-grâce and, although it may not be pleasant to human eyes, we cannot but admire the efficiency with which the kill is made. The lightning strike of a snake has a certain finesse about it and there is something spectacular, even majestic, about the final leap of a leopard or a tiger as it brings down the prey.

Such qualities cannot be attributed, however, to many of the invertebrate predators with slow-moving prey. We have already seen how the ladybird bumps into a victim and simply chews it alive. Some of the carnivorous snails (see page 79) have even more unpleasant habits, and there is certainly no finesse about the harvestman; detecting a likely victim, such as a small insect, with its long, sensory second pair of legs, the harvestman simply falls on it with what has been described as a 'pile-driver' action and begins to eat it. If it struggles, the victim is held down by the spider's long legs until it is finally overpowered.

Assorted Assassins

As we have seen earlier in this book, an animal does not have to be big and fast to be a killer. In fact, some of the most fascinating predatory behaviour is found in the slow-moving invertebrates, and these have certainly evolved some remarkable equipment. Pride of place for weaponry perhaps goes to the sea anemones and their relatives (see pages 38 to 43), but the terrestrial invertebrates exhibit no less enterprising methods of capturing their food.

Assassin bugs

Assassin bugs, which belong to the family Reduviidae, get their name from the way in which many of the species pick out another insect as a victim and then spear it to death with a powerful beak, or rostrum. This beak is like a hypodermic needle and it is possessed by all bugs, both plant-feeders and predatory species. Among the 3,000 or so species of assassin bugs it is particularly large and strongly curved.

The assassin bugs have long legs and many species can run down their prey. The front legs are powerful and often very bristly and, although not grasping to the degree found among the mantises (see page 30), they can fold round and hold quite large insects. Having caught its victim, the bug then plunges its beak into it, selecting a weak spot if the prey is a heavily armoured species such as a beetle. Toxic digestive juices are then pumped into the prey. One African species about 4cm (1·6in) long can even subdue the rhinoceros beetle several times its own size.

One constituent of the saliva attacks the nerves, and the prey is paralyzed in seconds. Another constituent then gets to work to digest the muscles and other body organs of the prey. Like all bugs, the assassins can ingest only liquid material, and they cannot feed on their prey until the digestive juices have liquefied its body contents. Most bugs have two slender channels in the beak — one through which the saliva is injected, and one through which the digested material is sucked up — but the assassins have just one large channel, enabling them to pump in large quantities of saliva and thus to subdue prey much larger than themselves. A large victim will provide food for several days, and the bug can be seen sitting quietly sipping for the whole period, during which its body becomes considerably distended. At the end of the meal, there is nothing left of the prey but a horny skin.

Some assassin bugs display aggressive mimicry, in that they resemble the gnats and other small flies on which they feed and thus are able to mingle with their prey undetected. Although most assas-

sins are true hunters, some wait in ambush for their food and may even lure the prey to its death. One species living in the West Indies secretes a sweet fluid that attracts ants. The ants flock to drink, but they become intoxicated and fall easy prey to the bug. Other assassins dip their legs into resin oozing from trees, and then use their legs like fly-papers to catch unwary insects.

The toxic saliva of the assassin bugs can be fired from the beak of some species to act as a protection against birds. It can also produce intense pain when an assassin bites man or another large animal. This is not unusual because some assassins have abandoned the insect-eating habit and now feed by sucking vertebrate blood. The actual blood-sucking is not serious as far as man is concerned, although the wound may be very painful, but the real danger is that the bugs can transmit a number of dangerous diseases.

Assassination
Above Undeterred by the warning colours and the unpleasant taste, an assassin bug plunges its beak into a ladybird and prepares to suck the victim dry.

Death in the dark
Right Using its eyes in the dim light, a long-legged centipede has captured a cave-cricket with its strong fangs. The prey is now being torn apart and eaten.

Voracious centipedes

The centipedes are all primarily carnivorous creatures and all are provided with a pair of venomous fangs. These arise just behind the head and they are often so large that they curve right round the head and almost meet in front of it. The poison gland is situated part way along the fang and a slender canal carries the poison to the tip, from where it is injected into the prey.

Few temperate species are big enough to hurt people — their fangs are not normally strong enough to pierce human skin, and if they do get through the amount of poison injected is not enough to cause more than local pain and swelling. The tropical species, however, are a different matter. Some of them reach lengths of 30cm (12in) and they feed on small rodents, birds, and lizards. Their powerful fangs and venoms can inflict painful wounds on humans, although they are probably not dangerous to healthy adults.

Most centipedes live in the soil and among decaying leaves and they feed on small insects, slugs, and other centipedes. Their food is detected mainly by touch, using long hairs on the antennae and other parts of the body. It is then usually grabbed by the fangs, although some small centipedes of the genus *Cryptops* can also capture prey in their greatly enlarged hind legs, which snap shut around the prey like the hind legs of *Bittacus* (see page 29). One group, typified by the house centipede (*Scutigera coleoptrata*), rely mainly on sight to capture fast-moving prey. They have very long legs and they move incredibly fast over walls and even ceilings as they dart after small flies. Their eyes are much better developed than those of other types of centipedes.

Most of the larger centipedes also rely partly on sight. In tropical countries they can sometimes be seen gathering around lights like geckoes (see page 88), ready to pounce on the numerous moths and other night-flying insects that are attracted after dark. Professor Cloudsley-Thompson even describes one captive centipede (*Scolopendra cingulata*) that reared up to snatch bees and wasps.

After the capture, the prey is torn to pieces by the mouthparts while it is held firmly by the fangs. The body fluids and other soft parts are ingested, while the harder parts are discarded.

Hooking the worm
Below The slug *Testacella* hooks a worm with its toothed radula (left). Withdrawal of the radula (centre) draws in the prey; extension (right) pulls in the rest.

Trapped in the jaws of death
Above The small and soft-bodied larva of the lacewing fly looks harmless enough, but it is a voracious destroyer of aphids and many other small insects. Here it has caught a young beetle larva in its big, sickle-shaped jaws. Each jaw, or mandible, is grooved on its lower surface, and a lobe of the maxilla — another of the mouthparts — fits into the groove to form a tube. The two jaws are plunged deeply into the prey, whose juices are then sucked up through the tubes. Many larvae camouflage themselves with prey skins.

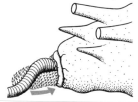 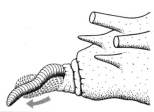

Oesophagus
Oesophagus pouch

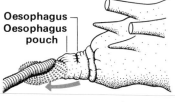

Worm-eating slugs

Most slugs are vegetarians or detritus eaters, but members of the genus *Testacella* have adopted a carnivorous diet and they feed almost entirely on earthworms and other slugs. These slugs, which are not uncommon in Europe, are easily recognized by their long, tapering bodies with a tiny saucer-like shell on their back. This shape enables the slugs to follow worms along their burrows. The head end — the narrower end — is equipped with a long tongue, or radula, clothed with needle-like teeth. These point backwards and, with the radula extended, they impale one end of the worm. Withdrawal of the radula into the head then draws the worm smoothly into the slug's voluminous mouth, much as we might suck in a string of spaghetti.

Exploding snails

A number of large snails, including *Paryphanta* of Australia and New Zealand,

also eat earthworms in the manner of *Testacella*, but most carnivorous snails feed on slugs and other snails. Some actually use their radula to drill through the shells of their prey (see also page 174), but others take an easier route, following the slime trails and attacking the exposed tail end of the prey. The prey is thus eaten alive, with the predator ending up by delving deeply into the shell to reach the concealed parts.

The Australian *Strangesta capillaris* has a very strange method of dealing with young garden snails (*Helix aspersa*). It enters the mouth of the shell and then expands the front end of its body like a balloon. The pressure generated is enough to burst the shell of the prey, thus exposing the flesh for the predator. Adult prey is eaten in the more 'conventional' manner described above — presumably the adult's shell is too strong to be exploded by pressure alone.

The Formidable Hunting Spiders

We have already met spiders that ambush their prey and spiders that make elaborate and highly efficient traps, but we have by no means finished with the spiders. Many species are out and out hunters, some nocturnal some diurnal, some fast and some slow, but all remarkably efficient at seeking out and running down their prey. Most of the wolf spiders (page 82) are nomadic, settling down wherever they happen to find themselves after a day of hunting and sunbathing, but the other hunting spiders generally have a permanent silken retreat to which they return after each expedition.

Hunting spiders take a very wide range of food, from the birds and small mammals taken by the tropical bird-eating spiders, to woodlice, ants, and other spiders. Many species have very catholic tastes and eat almost anything that they can catch and overpower, but others are more fussy and restrict themselves to particular groups of prey.

The woodlouse-catcher
Several kinds of spider can be found lurking in the woodpile or under a heap of stones, but one stands out from the rest by its brick-red front end and its huge jaws. This is *Dysdera*, the powerful killer of woodlice.

The slow, groping gait and poor eyesight of this nocturnal spider do not hinder its predatory activities, for its preferred prey are equally slow-moving. Most spiders are unable to pierce the hard exoskeleton of the woodlouse, and even those that can pierce it generally reject the woodlouse because of its unpleasant flavour. But *Dysdera* is a specialist and its enormous jaws are not without function. As soon as *Dysdera* bumps into a woodlouse of an acceptable kind – some are repellent even to the woodlouse-catcher – and picks up its scent, the huge fangs are unfolded and, with a slight twist of the spider's body, one fang is thrust under the woodlouse and into its soft belly while the other is sunk into the dorsal surface.

Drassodes – a formidable foe
Often found together with *Dysdera* are the mouse-coloured *Drassodes lapidosus* and its relatives. *Drassodes*, which also lives in drier places, such as grass tussocks, lacks the huge jaws of *Dysdera*, but it is nevertheless a formidable spider of the night. W. S. Bristowe, who has studied the feeding and mating habits of so many spiders, stated that he would 'back this ferocious gladiator against any other hunting spider with the possible exception of the large-jawed *Dysdera*'.

Drassodes prowls about on the ground with its front legs stretched out in front

of the body like antennae, and as soon as it bumps into another animal these legs seem to tell it all it needs to know. If it meets a small creature it immediately strikes with its fangs, but larger creatures are attacked in a very different way, especially if they are large spiders.

With a surprising burst of speed, *Drassodes* darts towards its adversary and then veers slightly to one side or even springs over the top, trailing a wide band of silk as it goes. The silk holds down the legs or body of the victim, allowing *Drassodes* to come straight back in for the kill – a deep bite into the victim's back.

Courtship is clearly a great problem for the male *Drassodes*; he must convince a mate that he is a suitor, not just food. The same difficulty is, of course, experienced by other male spiders, but *Drassodes* solves it in an unusual way. He lines up a mate before she undergoes her final moult and then, without any preliminaries, mates with her as soon as she finishes moulting. With her exoskeleton still soft, she poses little threat to the male, but at other times she is likely to regard him as food rather than as a mate, and the male must beware.

Ero the pirate
Spiders of the genus *Ero*, together with other members of the family Mimetidae, are sometimes referred to as pirate spiders because of the way in which they raid the webs of other spiders. Unlike human pirates, however, they are not interested in looting the webs; they are murderers intent only on killing and eating the rightful occupants of the webs.

Ero commonly invades the scaffold-type webs of *Theridion* species (see page 52) or the hammock webs of *Linyphia* species (see page 50), and again Bristowe has shown just how the slow-moving *Ero* deals with its more agile victims. It may wait until the victim comes within range of its own accord, but if this does not happen *Ero* gives the web a few gentle jerks. This quickly brings the web-owner down to investigate, and *Ero* strikes, snatching the victim in its long, hairy front legs and sinking its fangs into one of the spider's legs.

The victim dies almost instantaneously from what must be a very potent venom, but it is interesting to note that death is very much slower if the victim is bitten on the abdomen. A money spider bitten there may still be alive after 30 seconds.

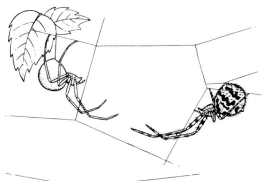

The poisonous adversary

Above Despite its slow movements, *Ero* is a deadly adversary of other spiders such as *Theridion*. Sitting motionless in its own flimsy web, or more often invading the *Theridion* web, *Ero* raises its long front legs and waits for *Theridion* to come within range. Drawing its victim in close with its front legs, *Ero* plunges its fangs into one of the prey's legs and injects a very fast-acting poison which kills almost at once.

Armour penetration

Above Dysdera has enormous fangs to pierce the tank-like armour of its prey. One has pierced the top of the woodlouse and the other will be plunged into the belly.

Death of a fly

Right Drassodes trails silken threads as she stalks her quarry. Pouncing with lightning speed, she grabs her prey and inserts her poisonous fangs deep into its neck.

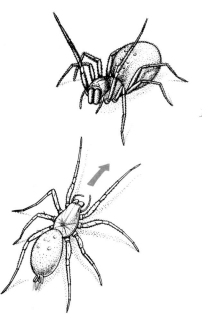

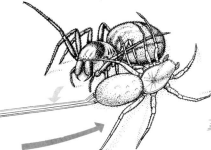

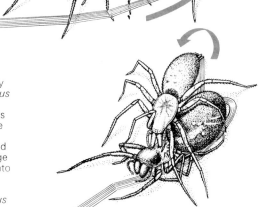

Drassodes straight-jackets its victim

The sleek, pale brown *Drassodes* fearlessly tackles larger spiders, such as the *Amaurobius* shown here, with the aid of its silken threads. With front legs raised, *Drassodes* darts around its prey, immobilizing its legs with silk. The broad band of silk is laid over the legs and the tip of the abdomen before being securely anchored. *Amaurobius* snaps its large and powerful fangs, but is powerless to retaliate in the face of this speedy onslaught and *Drassodes* can plunge its own long fangs into its victim's back. Its highly toxic venom paralyzes *Amaurobius* almost at once.

Th
its
al
ch
by
er
al
ra
le
of
rig
w

ti
ju
m
sp

Frogs and Toads

The frogs and toads, together with the newts and salamanders, belong to the class of vertebrates known as the amphibians. Most of them spend their early lives in water (as tadpoles) and their adult stage on land, although there are a number of completely aquatic species and some entirely terrestrial ones. The life cycles of the terrestrial species have become modified to cut out the free-living tadpole stage; the entire tadpole phase is passed inside the egg, which hatches to release a miniature air-breathing adult. The eggs are usually laid in damp ground, but some species actually retain their eggs inside the female's body until they hatch. The females thus give birth to fully-formed, miniature frogs and toads.

Useful predators

This section deals only with those frogs and toads that spend all or most of their adult lives on land. These animals are all carnivores and they eat a very wide range of other creatures, ranging from worms and slugs, through all kinds of insects and spiders, to mice and birds. They are very useful destroyers of insect pests, and large toads are even sold for pest control in various parts of the world. *Bufo marinus*, a giant toad from South America, was taken to Puerto Rico in the 1920s in an attempt to control the chafer beetle grubs that were destroying the sugar cane. This project met with some success, although the environment was not ideal for the toads and they gradually died out.

Some of the large frogs and toads reach total lengths of about 45cm (18in) and can inflict painful bites with their large jaws, but none is venomous in the accepted sense of the word. They rely on their strength to overcome prey and do not poison it in any way. The arrow-poison frogs and the various poisonous toads have poison glands in their skins, but these are used as a defence against predators and not as offensive weapons.

Ambushers or hunters?

Many frogs and toads could equally well have been dealt with in the section on ambushing, for they certainly seem to spend a lot of time just sitting around and waiting for food to appear. Nevertheless, they will go in search of food when they are really hungry, and some species are remarkably active in the pursuit of prey. The common toad (*Bufo bufo*) is a true hunter, and it has a distinct hunting territory up to about 450m² (4,842ft²). It seems to know its way around the area and it will return again and again to those parts of the territory where it has been successful in catching prey with ease in the past.

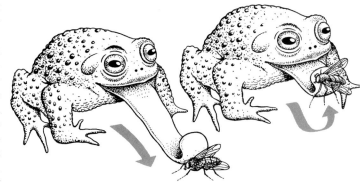

The sticky trap
Left The protruding, jewel-like eyes of the toad give a wide view and locate any prey within reach of the sticky tongue. The toad lunges forward and shoots outs its long tongue, which is hinged at the front to increase its range. The tongue brushes the roof of the mouth as it flips forward and picks up a coating of mucus.

Trapped by the tongue

With very few exceptions, prey is detected entirely by sight, the bulging eyes providing all-round vision in colour. Movement provides the essential trigger for hunting behaviour, and the anurans – the technical name for the frogs and toads – do not normally take any notice of dead food. When potential prey has been sighted, the frog or toad turns towards it and gets into position for an attack, often creeping stealthily forward by moving one leg at a time. If the prey is a slow-moving creature, such as a slug, the predator may get very close and then use only its remarkable tongue to effect the capture.

The tongue is normally fixed at the front of the mouth, with the free end lying back in the throat, but it can be flicked out with amazing speed. As it is ejected, the tip brushes against a patch of glandular tissue in the roof of the mouth and it becomes coated with a very sticky mucus. It can then pick up any small animal that it touches. The common toad has a tongue about 2cm (0·8in) long which can be flicked out and brought back with prey in just over one tenth of a second. The American bullfrog (*Rana catesbiana*) is even quicker, and its long tongue actually wraps right round its prey.

Even very active prey, such as flies, can be caught without difficulty at these speeds, and if the frog or toad cannot get near enough simply to flick out its tongue it may launch itself into the air

Quite a mouthful
Left The brown frog (*Rana wittei*) has caught a large bush-cricket that seems almost too big for it but, using its front feet to push in the legs and antennae, the frog will eventually engulf its prize. Frogs are very tenacious creatures and rarely let go once they have a grip.

Frog eats toad
Left The South African bullfrog seen here is actually a giant toad. It does not hesitate to snap up smaller toads in its massive jaws. Small, pointed teeth around the jaws and on the roof of the mouth prevent the luckless prey from struggling free, and it is eventually swallowed whole.

Taking the worm
Below The common frog (*Rana temporaria*) eats almost any small animal that it can catch, but it seems particularly fond of juicy earthworms, which are easily caught when they come to the surface at night. They are snapped up and swallowed whole by being sucked down.

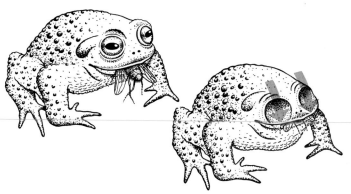

Blink and swallow
Left Frogs and toads blink when swallowing prey. This action pulls the large eyeballs down into the head, and the base of each eyeball bulges into the mouth, where it helps to force the food down into the gullet and through into the stomach. Swallowing is slow, and even a moth may take several minutes to get down.

and snatch the prey as it sails by. Many frogs and toads actually leap out of the water to catch flying insects in this way, and the tree frogs regularly leap at their prey with great success.

Not all the anurans have protrusible tongues, but the tongue is nearly always sticky and even if it reaches only to the lips, it can help to trap small prey as the frog or toad lunges forward with its jaws. Larger prey is caught with the jaws alone, and held by small teeth or sharp ridges on the jaw margins and in the roof of the mouth.

The front legs may be used to cram large or particularly active victims into the mouth. Long-legged spiders and large-winged insects are especially troublesome in this respect. The fingers of the front limbs may also be used to clean slugs and worms, the amphibian drawing the prey through the fingers into its mouth to scrape off mud and other debris.

Little or no chewing takes place, and the food is generally forced down the gullet in one piece. Strangely enough, the eyes help in swallowing; muscular action pulls the eyeballs down into the mouth cavity, where they help to force the prey into the throat. Swallowing large prey may take some time, but the anurans have no difficulty with their

breathing because a large amount of their oxygen is absorbed through the thin skin.

Most frogs and toads are rather catholic in their feeding habits, taking more or less anything that presents itself, but some species have become specialists. The Mexican burrowing toad (*Rhinophrynus dorsalis*), for example, specializes in eating termites, and it has a conventional type of tongue with which it laps up the small insects just like the other termite and ant-eating predators described and illustrated on page 118.

The taste test
Although prey is generally detected by sight, and the anurans usually lunge at or flick their tongue at anything of a

suitable size, other senses are involved in determining the acceptability of the prey. Some worms, for example, give off unpleasant odours and frogs and toads refuse to attack them. Most other prey has to be tasted to determine its palatability, and if it is distasteful it is ejected from the mouth almost as quickly as it is taken in. Many experiments have been conducted on the acceptability of various insects, and they have shown that the frogs and toads, like most birds and lizards, are able to appreciate the significance of warning coloration. Most unpalatable insects are brightly or boldly coloured, and when once the frogs and toads have tried a few of them they learn to leave them alone, even when they are really hungry.

The Versatile Lizards

The lizards are the most diverse and adaptable of all the living reptiles. Ranging from about 3cm (1·2in) to nearly 3m (10ft) in length, they include legless burrowers and tree-top gliders as well as the more familiar fast-running species that live on the ground. Their feeding habits are equally diverse. The majority are insectivorous, but many other animals are eaten by the world's 3,000 or so species of lizards. Some of the larger lizards are herbivorous, and it has been suggested that these plant-eating species have been able to attain large sizes simply because they expend little energy in finding food.

The Komodo dragon
The largest lizard of all is an undoubted flesh-eater. It is the Komodo dragon (*Varanus komodoensis*), a giant monitor lizard living on the island of Komodo and a few neighbouring islands in Indonesia. A stout, crocodile-like creature, it reaches a length of nearly 3m (10ft) and weighs about 140kg (308lb). Much of its food consists of carrion, which it seeks out with the aid of its forked tongue. As in the snakes and most lizards, the tongue picks up scent particles from the ground or the air and carries them back to Jacobson's organ in the roof of the mouth (see page 91).

But the Komodo dragon does not exist entirely on carrion; adults will kill deer, monkeys, and various other animals with their powerful clawed feet. The claws and the serrated, saw-like cheek teeth tear and slice the flesh, which is then gulped down in large chunks. The dragons have huge appetites, and sometimes eat so much at a sitting that they are unable to move far for several days.

Young Komodo dragons will feed on

The living dragons of Indonesia
Above The giant flesh-eating lizards of Komodo and other Indonesian islands appear as ferocious as any mythical beast. Despite their lumbering appearance, they can run with considerable agility to capture prey, killing it with their powerful claws.

carrion if they get the chance — the adults generally push them out of the way when they meet — but they more frequently eat birds, rodents, insects, and other lizards.

Poisonous lizards
Lizards generally overcome their prey by their superior size and strength, and we do not find any of the elaborate venom apparatus such as is found in many snakes (see page 92), but two species do have poisonous bites. These are the Gila monster and the beaded lizard, two closely related species of *Heloderma* from the deserts of Arizona and neighbouring areas. The venom glands are in the lower jaw, but they are not connected to the teeth and the venom simply flows into the mouth cavity. Some of it then gets into the victim's body when the lizard bites, but it is doubtful whether the poison plays much part in subduing the prey.

These lizards feed largely on eggs and on nestling birds and new-born rodents, none of which has much chance against the strong jaws even without venom. It seems that these two lizards simply have poisonous saliva, as do many snakes and, reputedly, some mammals. The Gila monster and the beaded lizard both spend much of the dry season in their burrows, and rarely come out to feed, even at night. During this period of fasting, they exist on food reserves accumulated during the previous rainy season. Much of the food is stored in the tail, which becomes very

fat before fasting starts. In captivity, the Gila monster has been known to survive without food for three years.

Speed of foot
Anyone who has tried to catch, or even to photograph a lizard on a wall or on the ground knows that the typical lizards are extremely quick and agile. This enables them to escape from many of their enemies, and it also helps them to catch their own prey. Even the nimble grasshopper frequently falls victim to the lightning dash of a lizard. A combination of scent — picked up by the forked tongue and detected by Jacobson's organ — and sight is used to find prey, and the lizard then darts forward to grasp the food in its jaws.

Prowlers on the ceiling
Geckoes are a fascinating group of lizards with remarkable adhesive pads

The fly-catcher
Above The gecko breaks its camouflage cover as it lunges at its prey. It can adhere to almost any surface by rolling down its flattened toes. To lift its foot the lizard curls up its toes and peels off the ridged pads (*right*) with microscopic hairs.

on their toes. They are beautifully adapted for hunting on rocks, tree trunks, and other vertical surfaces, although a few species have put their broad toes to good use on sand dunes. House walls are perfectly acceptable to the rock and tree-dwelling species, and their adhesive toes even allow them to run across the ceilings.

Each toe has a number of overlapping flaps, or lamellae, on the underside, and these are clothed with microscopic, backward-pointing hooks. The hooks are so small that they can get a grip on the tiniest irregularity, and their combined effect is more than enough to hold the gecko on a ceiling or even on a window pane.

Most geckoes are active at night, and they rely almost entirely on their superb eyesight to catch their insect prey. Scent may help them to some extent, but they do not rely on Jacobson's organ in the way that many other lizards do, for the tongue is much thicker and it is not continually being thrust out of the mouth to pick up scent particles.

When suitable prey is seen, the gecko moves stealthily into position and then makes a lightning dash to grab it in its jaws. The geckoes soon learn that abundant prey can be found around lights on walls and ceilings, and large numbers of these lizards congregate in such places at night. They are welcome guests in most tropical houses because of the huge numbers of flies that they destroy with their darting technique.

The amazing chameleons

The chameleons are perhaps the most famous of all the lizards, partly for the way in which they can change their colours to match different backgrounds and partly for their amazing fly-catching abilities. They have evolved along totally different lines from other lizards, becoming very slow, arboreal creatures and developing the tongue into a mobile 'fly-paper' even more efficient than that of the toads. A chameleon 18cm (7in) long may have a tongue reaching 30cm (12in) in length when extended, and the speed and accuracy with which it operates is truly amazing.

Accuracy is controlled by the eyes, which are the sole means of detecting prey – Jacobson's organ being virtually absent because the tongue no longer acts as a scent-collecting organ. The eyes are housed in turrets and they can be moved independently of each other: one can look backwards while the other looks to the front. This means that the animal can scan a wide field for food, but when an insect is detected the chameleon can turn and bring both eyes to focus on it at the front, the binocular vision giving accurate range-finding to ensure a successful strike.

Firing the tongue

The chameleon's tongue is an extremely elastic body, and it is hollow for much of its length. The tip is solid and covered with glands producing a very sticky secretion. When the animal is at rest, the 'stalk' of the tongue is folded up like a concertina, with a bony outgrowth from the floor of the mouth fitting neatly into the cavity like a core.

When prey is sighted within range of the tongue, the bony core is moved forward to bring the tip of the tongue right up against the lips. A sudden contraction of the circular muscles in the stalk then causes it to elongate and send the tip shooting out with an action that has been likened to that of spitting out an orange pip. The prey is

Shooting the prey
Above Once its independently rotating eyes are fixed on the insect, the camouflaged chameleon stalks slowly and cautiously within range. Then with lightning speed it shoots out its long, elastic tongue, hitting the target with the sticky tip. The victim is gummed to the tip and pulled back into the chameleon's snapping jaws. After a few perfunctory chews, the prey, such as the locust shown here, is usually swallowed in one gulp.

Extending tongue
Right The elastic tongue is folded round a long thin bone in the floor of the mouth. On extension, this bone is levered forwards by other bones and muscles. Circular muscles in the tongue contract, shooting the sticky tip forwards. The tongue can extend to almost twice the length of the body.

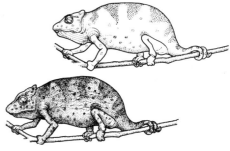

The dappled camouflage of the chameleon
Above The chameleon can alter its colour to match the foliage.
Below The changes are produced by the migration of pigments.

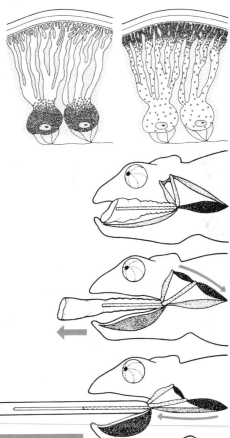

trapped by the gummy tip and the whole tongue is retracted by the elastic fibres and a second set of muscles – all within the incredibly short space of about 40 milliseconds.

Sniper in camouflage

The chameleon's excellent all-round vision and its long tongue enable it to see and catch a lot of insects without moving, other than to turn round to face the prey. In this respect, the chameleon is an ambusher, and it is greatly helped by its colour-changing abilities. It cannot match every background, but, depending on its 'basic' colour, it can certainly blend in with a range of green, brown, and yellow backgrounds, thus providing good camouflage in its natural habitat among the trees. The shape of the body, and the arrangement of the various bands and blotches of colour

on it also help with concealment. Some species are extremely flattened from side to side and incredibly leaf-like. They even sway gently on the twigs and leaves.

Despite these adaptations for ambushing, however, the chameleons do go after their prey when they see something that is outside striking distance. Gripping the branches with their specially modified feet and prehensile tails (unique among lizards), they move slowly forward with their eyes glued firmly on the prey until they are within striking distance. Chameleons rarely let go with more than one foot at a time, and they always get themselves rock-steady before firing – with the result that they rarely miss. The prey may escape before the chameleon has got into position, but it has little chance of avoiding capture when once the bullet-like tongue has been fired towards it.

ground, and when it is withdrawn into the mouth it carries scent particles to the sensory tissues in Jacobson's organ. The snake thus picks up a very detailed scent picture of its surroundings. Many snakes rely entirely on the sense of smell to track down their food, although some movement on the part of the prey may be necessary to trigger off the final strike.

The rattlesnakes and pit vipers, including the bushmaster (*Lachesis muta*) of South America and the copperhead, possess remarkably efficient heat sensors in pits situated between the eyes and the nostrils. The pits are lined with delicate membranes and numerous nerve endings which are so sensitive to heat that the snakes can detect temperature changes as small as $0.003\,°C$ $(0.005\,°F)$. This amazing sensitivity enables the snakes to detect the presence of a warm-blooded animal some distance away and to home in on it and strike very accurately even in total darkness. Some boas also have heat-sensitive pits on the scales fringing their lips, although these pits are less sensitive and efficient than those of the pit vipers.

Catch and kill

Having no limbs, the snake's only real weapons are its teeth and jaws. Whether the prey has been ambushed or stalked, the initial attack is always a rapid lunge and a bite. The speed of the strike has been timed at about 3.5m (11.5ft) per second in the prairie rattlesnake (*Crotalus viridis*). This is not as fast as a man can punch, and much slower than one might imagine, but fast enough to take the snake's victim by surprise. The next stage of the proceedings depends on the type of snake and also on the size of the prey.

The European grass snake (*Natrix natrix*) displays some of the simplest feeding behaviour. This species feeds on frogs and other small animals with little or no defensive equipment. The snake does not let go when once it has bitten the prey and it then swallows it alive. Many other small snakes feed in the same way. Snakes that deal with larger prey, or prey which struggles or has sharp teeth or claws with which to defend itself, must kill their victims before attempting to swallow them. The killing is done by constriction or by poisoning.

The constrictors

Constricting snakes use the muscular power of their bodies to subdue their prey, usually killing it by suffocation. Contrary to popular thought, they do not crush their victims: they merely exert enough pressure to stop them from breathing. All the really big snakes belong to this group – the pythons, the

Muscle power
Left An Indian python has coiled its body around a hog deer and is squeezing it to death. The snake merely exerts enough pressure to prevent the prey from breathing. Such a meal will last the snake several weeks.

Varied dentition
Right The jaws of four kinds of snake showing the main types of tooth arrangement. The boas, pythons, and other non-poisonous snakes (1) have teeth that are all much the same size. Poisonous snakes have some teeth enlarged to form fangs. The back-fanged snakes, such as the boomslang (2), have fangs near the rear of the mouth. Elapid snakes, such as the mamba (3), have permanently erect fangs at the front of the mouth. Vipers (4) have huge fangs that are folded up when not in use and erected when the mouth opens.

boas, and the giant anaconda, which is generally regarded as the largest of them all. It reaches lengths of more than 9m (29.5ft) and it can overpower, and eat, caymans and bulky animals such as deer, pigs, and capybaras. The large pythons also feed regularly on deer and pigs.

Although one immediately thinks of the big snakes when constriction is mentioned, this method of killing prey is in no way restricted to the large species. Many of the slender, snake-killing species use constriction. All the constrictors are well armed with teeth and can give painful bites, but they are not venomous.

The typical constrictor may ambush its prey or track it down by scent. It then strikes when the prey is within range.

Boomslang – back-fanged, but dangerous
Above The African boomslang snake, the most dangerous of the back-fanged species because of the rather forward position of its fangs and its unusually toxic venom, spends most of its time in the trees, where it feeds largely on chameleons, birds, and tree frogs.

Even the apparently sluggish pythons can strike remarkably quickly. The teeth are sunk into the victim and, before it has time to react, it is encircled by one or more coils of the snake's body. The teeth release their hold, and the body applies pressure steadily. Every time the victim breathes out, the snake tightens its grip a little more. Breathing in becomes more and more difficult, and eventually impossible. The prey is dead, and the snake releases its hold and begins to examine its meal with its tongue. The examination may be thorough and time-consuming, but it serves to establish which is the head end, at which the snake must begin the long-drawn-out business of swallowing its victim. The snake's backward-pointing teeth are ideal for holding its prey, but they are useless for cutting or chewing and the snake has no alternative to swallowing the prey whole.

The poisoners

Many small snakes considered harmless to man actually have slightly poisonous saliva, which may help to subdue their prey even if it is not strictly injected. The truly venomous snakes have greatly elaborated this arrangement, with some of the salivary glands being transformed into true poison glands and some of the teeth being modified to inject the poison efficiently into the prey. The glands occupy a considerable part of the head, and sometimes extend well back into the body as well. The Malaysian banded coral snake (*Maticora intestinalis*) has venom glands that extend almost a quarter of the length of its body.

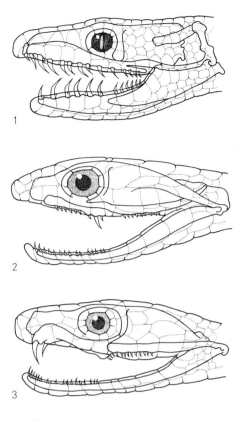

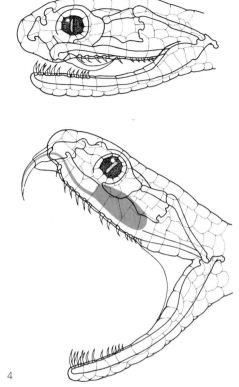

The non-poisonous snakes generally have all their teeth of a similar size, but the venomous species have some teeth which are much larger than the others. These large teeth are the fangs. The family Colubridae contains about half of the known snake species and the great majority of them are non-poisonous. The few venomous species, including the African boomslang (*Dispholidus typus*), have their rear-teeth modified as fangs and are therefore known as back-fanged snakes. The fangs carry slender grooves on the hind margin, and the venom flows gently into the wounds rather than being pumped into them.

Apart from the boomslang, few back-fanged snakes are dangerous to humans. They cannot open their mouths wide enough to bring the fangs into action against a large animal, and even when catching their normal small prey, such as mice, frogs, and birds, they have to hold tight with the first bite and 'chew' vigorously in order to dig the fangs in and get enough venom into their victims. The boomslang is dangerous because its fangs are situated much further forward in the mouth than they are in other back-fanged snakes, and they can strike directly into quite large animals. The snake is also unusual in that it has three pairs of fangs instead of the usual one functional pair. Its venom is more powerful than that of the cobra, but relatively little is produced and human deaths are rare.

The family Elapidae contains the cobras, mambas, kraits, coral snakes, and sea snakes. These reptiles produce some extremely potent venoms and

cause many human deaths, but even so they are not quite as dangerous as they are painted: most of them are shy and retiring animals and strike only when trodden on or otherwise provoked. The single pair of fangs are situated at the front of the upper jaw and, although quite short, they stab directly into the victim when the snake strikes. The fangs are tubular and connected directly to the venom gland. Venom is pumped straight through the fangs and into the wounds by the contraction of muscles around the venom gland.

The largest of the poisonous snakes belongs to the Elapid family. It is the king cobra, or hamadryad (*Ophiophagus hannah*), a snake-eating species that reaches lengths of about 5·5m (16·5ft). Its fangs are a mere 1·5cm (0·6in) long, but they

The largest snake fangs in the world
Above The gaboon viper possesses the largest of all snake fangs – up to 5cm (2in) long. The snake is one of a group known as puff adders and it hisses a loud warning when large animals approach. Extremely poisonous, its main prey are rodents and birds.

can deliver more venom than those of any other snake, and the venom is such that a bite from this snake can kill an elephant in four hours. Human deaths are rare, however, because the snake is not at all aggressive and it attacks only as a last resort. Far more deaths are caused by the much smaller, but commoner Indian cobra (*Naja naja*).

Another well-known species is the ringhals, or spitting cobra (*Hemachatus haemachatus*) from Africa. The terminal opening of the fang has moved round to the front surface, and the venom can be fired for distances in excess of 2m (6·5ft). This seems to be a defence mechanism against large animals that might either attack or simply tread on the snake. The venom is aimed at the eyes and causes temporary or even permanent blindness.

Elapid venoms work rather quickly when injected into the prey, and the snakes do not let go when once they have struck a suitable victim. They begin to swallow it as soon as it is still.

Elapid snakes are particularly common in Australia, where about 60 percent of all snakes are venomous and potentially dangerous. There are few colubrid snakes in the region, and none of the viper group, and so the elapids have radiated into nearly all the habitats. Among the well-known species are the taipan (*Oxyuranus scutellatus*), which reaches lengths of more than 3m (10ft), the tiger snake (*Notechis scutatus*), the brown snake (*Demansia textilis*), and the death adder (*Acanthophis antarcticus*). The latter looks and behaves remarkably like the puff adder of Africa, with a stout body, a broad, triangular head, and a sluggish nature, but it is not related to the adders or vipers. Its fangs show that it is a true elapid snake.

Vipers, pit-vipers, and rattlesnakes all belong to the family Viperidae. They are front-fanged snakes, but they are even more specialized than the elapids. Their fangs are very long – up to 5cm (2in) in the gaboon viper – and they inject large quantities of venom deep into their victims. When the snakes are at rest, the fangs are folded more or less flat along the roof of the mouth, but when the animals strike the fangs are rotated forward by movement of the jaw bones and they project well beyond the front of the mouth.

Although efficiently administered, the venom of the viperids tends to work more slowly than that of the elapids and most of the viperids release their hold as soon as they have injected the venom. They then wait for it to take effect before beginning to swallow the prey. The latter may wander some distance away before collapsing, but the vipers and rattlesnakes are nearly all ground-dwelling

creatures and they can easily find the victim again, even in the dark, with the aid of Jacobson's organ and, when present, the heat-sensitive pits.

Apart from the rattlesnakes of America, the best known viperids are the European viper (*Vipera berus*), which is the only snake known to live beyond the Arctic Circle, the gaboon viper (*Bitis gabonica*) and the puff adder (*B. arietans*) of Africa, the bushmaster of South America, and Russell's viper (*Vipera russelii*) of southern Asia. Those living in the cooler regions are active by day, but tropical species tend to hunt by night, although some forest-dwelling species may be active at any time. The gaboon viper, for example, tends to hunt at night and rest by day, when its superb disruptive coloration (see page 29) conceals it among the leaves, but it is often awake in the daytime and it ambushes quite a lot of its prey – mainly birds and rodents – during daylight hours.

New fangs for old
Snake teeth, like those of sharks and crocodiles, are continually being replaced. This is true of the fangs as well as the ordinary teeth. None of the teeth is particularly strong, and they frequently get broken during the struggle to capture and swallow prey, but the snake does not wait until it breaks a tooth before it renews it. New teeth are grow-

ing all the time alongside the ones they are to replace, and they are almost fully formed by the time the old ones fall out.

The fangs of elapids and viperids always grow in a sheath, and it is actually this sheath that is connected to the venom gland. Venom is pumped into the sheath and then into an opening at the base of the fang, from where it travels to the tip and into the wound. Because the venom flows first into the sheath, there is no problem in connecting up the venom duct and successive fangs.

Observations suggest that the rattlesnake fang does not last more than about ten weeks, and it may get broken much sooner. Early breakage does not speed up the replacement of the fang, but as long as the snake has one operational fang it can feed quite efficiently. Even if it should lose both fangs, it is not seriously handicapped, because all snakes can survive for long periods without food.

Complex venoms
Snake venoms are the most complex of all poisons. They contain several enzymes and other complicated proteins that destroy the tissues of the snakes' victims, although these enzymes are not necessary for the actual digestion of the prey. The make-up of the venom differs from species to species, but most venoms appear to contain four major groups of toxins. There is a component

that interferes with the nervous system and brings about paralysis or lack of co-ordination; there is a component that acts directly on the heart muscle and causes paralysis; a third constituent destroys the general body tissues; while a fourth acts on the blood of the prey, either causing massive clotting within the body – as is thought to happen when a viper bites a small mammal – or producing just the opposite effect of massive bleeding. Many snakes can control the amount of venom injected, and do not waste large amounts on small victims.

Swallowing the prey
A snake's prey often has a diameter considerably greater than that of the snake itself. This might appear to present an insuperable problem to a predator with no limbs and no cutting teeth with which to break up its prey, but the snake is equal to the task of swallowing such large prey whole, thanks to a number of special modifications of the skeleton and muscles of the head and jaws.

Before starting to swallow its prey, the snake almost always manipulates it so that it starts at the head end. Limbs fold down more easily this way when the snake reaches them, but even the snake-eating snakes start at the head end, which they detect by feeling the lie of the scales on their prey. Presumably the scales could scratch the

Egg-eating snake
Above left Dasypeltis,
the egg-eating snake,
shows its enormous
gape as it slowly
engulfs an egg. Only
when the egg is
completely engulfed
is the shell broken.

Slow-worm prey
Above The European
smooth snake, a non-
poisonous constrictor,
in the act of swallow-
ing a slow-worm.

Pushing in the prey
Right The European
ladder snakes uses the
hind end of its body to
help push a mouse
into its gaping mouth.

digestive tract if the snake were swallowed tail-first. Having found the head,
the snake may give it a good examination with its tongue and then, when
satisfied that it will make a good meal,
it makes a start.

The most obvious thing at this stage
is the snake's enormous gape — several
times wider than the head itself. This is
made possible by the extremely elastic
skin and also by the special hinging of
the lower jaw on the skull. In other
vertebrates there is just a single hinge,
but the snakes have two hinged joints.
The lower jaw itself is hinged to a bone
called the quadrate and can drop away
from the skull at this joint, but the

quadrate itself is hinged to the skull, and
it too can drop down, thus producing the
very wide gape. In addition, the bones of
the snout can be hinged up on the front
of the skull, and the left and right halves
of the lower jaw can be separated at the
front because they are joined merely by
an elastic ligament and not by fusion of
the bones themselves.

When the snake first takes a hold on
its prey with intent to swallow, it digs
in its top and bottom teeth and then
slides one half of the lower jaw forward
to obtain a new grip. The prey itself
remains firmly held by the rest of the
backward-pointing teeth. The other side
of the jaw is then eased forward, and

when it has a good grip the whole of the
upper jaw is pulled forward over the top
of the prey. The teeth all take a new grip
and the process is repeated again, and
again until the prey is completely engulfed. The constricting snakes often use
one or more coils of their bodies to push
against the prey and thus help it into
the mouth a little more easily. Copious
secretions of saliva are produced when
once the prey is in the mouth, and this
helps the later stages of swallowing.

The whole process can take several
hours, and another adaptation comes
into play to ensure that the snake can
breathe: the windpipe is pushed forward
over the tongue so that its opening is
right at the front of the mouth and freely
exposed to the air. Extra strong cartilaginous rings around it prevent its being
crushed. As the prey passes into the
throat, rhythmic muscular movements
take over and gradually pass it back
along the body. Flexible ribs and the
absence of a breast bone, combined with
the elastic skin, allow very large prey to
pass along the body.

Digestion is quite slow, especially in
the early stages, and the bump remains
in the body for quite some time. Large
constrictors that have swallowed pigs or
other bulky mammals remain swollen
and unable to move for days, and they
may spend weeks sleeping off the meal
before they feel the need to feed again.

The Badgers and Wolverine

The European badger (*Meles meles*) is one of the least carnivorous of the carnivores. It belongs to the same family as the highly predaceous stoats and weasels, but it has a great liking for fruit and other vegetable food. During the cold winter months the badger probably eats a great deal more plant material than animal matter.

Among those animals that it does eat, the most important are undoubtedly the earthworms. Their faint rustlings may be picked up by the badger's keen ears, or they may be sniffed out by its equally keen nose. It has been estimated that one badger may find and eat 200 earthworms in a single night. Most of these are likely to be caught within a period of about two hours – not a bad rate of working, even if most of the worms are discovered on the surface in damp weather. It is believed that the territorial boundaries of a badger colony are determined largely by the earthworm population of the area.

Other invertebrate food taken in appreciable quantity includes beetles – collected avidly from cow-pats and other dung – and slugs. The nests of wasps and bees are also raided quite frequently and, undeterred by the fury of the adult insects, the badger dines on the juicy grubs and, when a bees' nest has been attacked, the stored honey.

Ticklish and prickly problems
The badger's main vertebrate prey, at least in the British Isles, is the rabbit. Adult rabbits are rarely taken unless they are sick or injured, for they are far too quick for the badger, but young ones are frequently dug out from their nest-burrows, or stops.

The badger probably relies on its acute sense of smell to detect the rabbits under the ground, although it may hear them as well, and then it digs vertically down to reach them. It has not yet learned that there is an easier way in – through the original tunnel dug by the mother rabbit – and it relies on its instinctive vertical digging. The young rabbits are also quite vulnerable during their first week or two above ground, when they are still rather unwary and too slow to escape.

The badger swallows a good deal of fur when eating young rabbits, as shown by the amount of fur in its droppings, but it does not seem to relish the dense adult fur. Some adult fur is inevitably eaten, but the badger often skins the rabbit as it eats and leaves the skin turned neatly inside out as evidence of a good meal.

If rabbit fur is a ticklish problem for the badger, then hedgehog skin must present a prickly one. But it is a problem

The omnivorous European badger
Above The badger is a heavily-built and powerful animal, but it uses its strength for digging rather than for overpowering prey. It eats a lot of fruit and other vegetable matter, and a large number of earthworms.

that the badger certainly knows how to solve, for hedgehogs figure occasionally in the badger's diet. The badger attacks the underside of the hedgehog, where there are no spines, and leaves just the empty skin. It eats with such surgical precision that one badger whose stomach was opened for investigation was found to have eaten four hedgehogs and to have swallowed just three spines.

Rats, mice, voles, and moles are all eaten when the opportunity arises. The badger detects them by sound or by smell – its most important sense – and pounces upon them with its great paws. The sharp teeth make short work of the prey and very little evidence of the kill is left apart from the badger's footprints in the soil. Food is never carried back to the badger's burrow, or set, even when cubs are being weaned. The cubs take partly digested food regurgitated by their mother and then start hunting for worms and slugs on their own.

The American badger
The American badger (*Taxidea taxus*) resembles the European species except that it has a less boldly-striped head. Its behaviour is very different, however, for it is almost entirely carnivorous. It takes large numbers of voles and insects – especially bees and wasps – but its main prey are ground squirrels. These are taken more by an ambush technique than by active hunting, with the badger

lying in wait at the entrance to a squirrel's burrow until an animal emerges.

Badgers in the Cache National Forest in Utah were actually seen to block up all but one of the tunnels leading to a squirrel's nest and then to go in through the remaining entrance to make a kill. Such 'intelligent' behaviour has not been seen elsewhere and it seems likely that a small group of badgers learned to do it accidentally.

Birds and rattlesnakes also figure in the diet of the American badger which, unlike its European counterpart, caches uneaten food for another day.

The honey badger
The honey badger (*Mellivora capensis*), also called the ratel, lives in most parts of Africa south of the Sahara and extends eastwards into India. Although it belongs to the same family, it is not closely related to the European and American badgers. For its size, it must be one of the world's most ferocious animals; it stands only about 30cm (12in) high at the shoulder and it is no more than 75cm (30in) long, and yet it has been known to attack cattle and even the powerful Cape buffalo. One bold specimen was seen to tackle and kill a

Follow me
Right A honey-guide bird sees a ratel, or honey badger, and begins to call noisily and to flit from tree to tree. The ratel follows the bird and is eventually led to a bees' nest, which it digs out with its claws. The honey-guide is rewarded with a meal of beeswax when the ratel has eaten the honey.

Ratel meal-time
Below Although the ratel co-operates with the honey-guide, it is quite happy to eat other kinds of birds.

python more than 3m (10ft) long.

Powerful teeth and immense claws are the ratel's weapons, and its immensely thick and tough skin affords it a good deal of protection, even against the fangs of a cobra. For all its power, however, the ratel normally feeds on relatively small prey such as rodents, birds, and insects. Scorpions and tortoises are also eaten, the scorpion stings having no effect on the thick skin, while the tortoise shells prove no problem to the ratel's strong teeth.

As with the other badgers, most of the hunting is carried out at night, but the ratel does come out during the day and this is when it forms a remarkable relationship with a little bird known as the honey-guide. There are actually several species of these birds, of which the best known is *Indicator indicator*.

Showing the way
Like other badgers, the honey badger is extremely fond of honey and it frequently tears open bees' nests to get at it. It is quite capable of finding nests by itself, but in parts of Africa it may be assisted by the honey-guide.

The traditional story, based on scattered observations by Africans, is that the honey-guide finds a nest and then attracts the attention of a honey badger by flying about and giving a characteristic churring call. The badger follows the bird and is led to the bees' nest,

where it gets to work with its claws and gains a good meal. The bird gets its reward in the form of the beeswax, which it eats avidly.

Little serious investigation has been carried out into this strange behaviour, but Herbert Friedmann's work suggests that the true story might not be quite so simple. It seems more likely that the bird starts to call and flit from tree to tree when it sees a ratel, and this may be a form of mobbing behaviour. The chances are that if the ratel continues to follow the bird it will eventually be led to a bees' nest, but we do not know if the ratel's following response is instinctive or learned. The honey-guide's reaction to the ratel must be instinctive because the birds are brought up by foster parents, just like cuckoos, and do not associate with their true parents.

It is difficult to see what advantage the bird gets from guiding a ratel to a bees' nest, for the honey-guide is quite capable of entering a nest and taking beeswax and grubs for itself without getting hurt by the adult bees.

The gluttonous wolverine
The wolverine is the largest member of the weasel family, looking rather like a cross between a badger and a bear. It lives in the vast coniferous forests that spread around the northern parts of the world and is said to have an enormous appetite – hence its alternative name of glutton.

Many of the stories about its eating habits are probably greatly exaggerated, but there is no doubt that it is a powerful and courageous animal. Its main prey during the winter is the reindeer or caribou, although it generally attacks only the weaker animals, and it can drag carcases several times its own weight.

The wolverine also eats large amounts of carrion, a fact that has led to its being called the hyena of the north. The carrion is often in the form of animals caught in trappers' snares or the remains of a lynx or bear kill, but the wolverine is aggres-

Wolverine – hyena of the north
Above The heavily-built and powerful wolverine is often called the hyena of the north because of its scav-enging habits, but the animal is also an aggressive killer, tackling animals much larger than itself.

sive enough to drive both the lynx and bear from fresh kills as well. Its powerful jaws, like those of the hyena (see page 128), are able to crack the largest bones.

Despite its reputation for gluttony, the wolverine does not necessarily finish its meal at one sitting; leftovers are often buried in the snow for another day, although another wolverine is just as likely to find the cache as the one who actually buried it. These caches are an important source of food during very snowy weather, when the wolverine may not be able to hunt very easily.

The best hunting conditions are when the snow is just firm enough to bear the weight of the wolverine, but not hard enough to support a running reindeer. Even then, the wolverine probably catches most of its prey by leaping onto it from a rock or from an overhanging branch, for it is neither a speedy nor a stealthy hunter.

During the summer, the wolverine feeds largely on rodents, birds, and eggs, all of which it finds largely by night with the aid of its keen sense of smell. The animal is about during the daytime as well, however, and raids wasp nests for the grubs. In the autumn, it eats a considerable amount of fruit.

The Fast-Moving Weasels

The weasels and martens are lithe and fast-moving predators belonging to the family Mustelidae. Other members of the family include the stoat, the polecat, and the mink – all closely related to the weasels – the otters, the badgers, and the wolverine (see page 96). Some of these mammals, the martens included, eat appreciable amounts of vegetable food at certain seasons, but the weasels and other members of the genus *Mustela* are almost entirely carnivorous. They are specialist predators of small mammals and birds. The yellow-bellied weasel (*M. kathiath*) is sometimes kept as a 'mouser' in Nepalese houses, while the ferret – a domesticated race of the polecat (*M. putorius*) – has long been used for catching rabbits.

The weasels – small but deadly

The common weasel (*Mustela nivalis*) is widely distributed in Eurasia, and also in North America, where it is known as the least weasel to distinguish it from the rather similar long-tailed weasel (*M. frenata*). The male averages some 20cm (8in) in length, without the tail, and weighs about 100gm (3·5oz). The female is even smaller, and yet the weasel is really quite a ferocious creature and a formidable adversary for animals several times its size. Voles of the genus *Microtus* make up the bulk of its food, however – almost 50 percent in parts of Britain – with small birds, mice, rats, and a few rabbits contributing the rest. The long-tailed weasel eats rather more rabbits and rats, in conformity with its larger size.

Weasels are active by day and by night, and they use sight, smell and hearing to detect and follow their prey, although it is probable that they are first alerted by the slight rustling of small animals in the vegetation. The sense of smell is used to follow a trail and, being such a slim beast – it is said to be able to pass through a wedding ring – the weasel can follow voles right into their tunnels. The final attack consists of a rapid sprint and a leap, with the weasel clutching the victim in its front legs and aiming a deep bite at the back of the neck. This usually crushes the back of the skull and the prey dies at once.

Prey larger than the weasel, such as a rat or a small rabbit, is pulled on to its side so that it cannot run, and its back is pummelled by the weasel's hind feet as the predator struggles to make the neck bite. Movement is very important in triggering off the final assault, and a weasel getting into a hen house will often go on killing until there are no hens left, even though a single hen is more than enough for a meal.

A weasel eats only about 4gm (0·14oz) of food at one sitting, but it is a very active

The mink – water-loving relative of the weasel and the stoat
Above The American mink (*Mustela vison*) is equally at home on land and in the water and usually makes its home in river banks and in marshes. Prey captured in the water include frogs, crayfishes, and various species of fish, but the proportions vary a great deal from season to season. Other foods include small rodents and water birds.

The deadly neck-bite
Right A weasel sinks long canine teeth into the neck of a wood mouse – the typical way of killing its prey.

animal and it must eat regularly. Observations on captive animals suggest that they need to consume something over 25 percent of their body weight of food each day. This means some 20–30gm (0·7–1oz) every day and, allowing for wastage, this is equivalent to one or two voles – in good agreement with observations of wild animals that suggest a weasel eats about 500 voles or their equivalent each year. Uneaten prey is stored, often on a large scale, for use at a later date. It may be buried in a number of separate caches, or in one large larder in a central part of the weasel's territory.

Weasels normally hunt alone, but sometimes roam in pairs and occasionally in family groups, with the mother teaching her youngsters the basic hunting techniques. They have also been known to 'charm' young rabbits in much the same way as the fox (see page 102).

The stoat

The stoat (*Mustela erminea*) is a larger version of the weasel, with males averag-

The search for prey
Left A stoat stands up to look for prey. A rabbit (right) senses danger and sits up to look around as the stoat begins its stalk.

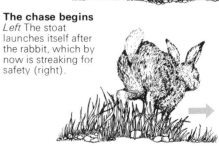

The chase begins
Left The stoat launches itself after the rabbit, which by now is streaking for safety (right).

ing 300gm (10·5oz) and females 200gm (7oz), although there are some much smaller geographical races in the southern United States. Like the weasel, it occupies a very wide range of habitats, from open tundra to dense forest. Its hunting methods are much the same as those of the weasel but, as one would expect, it concentrates on larger prey. In Britain it takes rabbits, small rodents, and birds (mainly game birds) in roughly equal proportions, often killing prey four or five times its own weight. Rather more rodents and fewer rabbits are eaten in North America.

Scent plays a major part in tracking down the prey, and the kill is made with a neck bite just as with the weasel. Each meal amounts to about 10gm (0·35oz), but the animals eat about one third of their own weight of prey every day and clearly return to a kill several times. There is no evidence of food storage by the stoat, which is very strange in view of the animal's close similarity to the weasel in other respects.

The martens
The martens, typified by the pine marten (*Martes martes*) in Europe and the American marten (*M. americana*) in North America, are much more omnivorous creatures than the weasels. Their diets show considerable seasonal variation and include significant quantities of insects and plant food. Both species live primarily in wooded country and are often described as important predators of squirrels, but observations on both sides of the Atlantic suggest that squirrels play only a minor part in the martens' nutrition. They are commonly eaten in some areas, but virtually ignored elsewhere, and it is possible that only certain populations of martens have acquired a taste for squirrel meat. All the martens are very agile and certainly able to catch sauirrels, both in the trees and on the ground, but voles and birds are their main foods in most places.

James Lockie studied the pine marten in Scotland and, from an analysis of its droppings, he concluded that small rodents and birds are the staple foods at all seasons, with an overwhelming preference for voles of the genus *Microtus*. Beetles, caterpillars, and berries are all eaten in season, and wasp nests are occasionally dug up. Some fishes and young rabbits are eaten, but few squirrels are caught in Scotland.

Male pine martens generally hunt alone, but the females commonly take their young with them. They are mainly nocturnal and they rely on the three major senses of sight, smell and hearing to find food. They may use aerial routes through the branches, although they catch most of their mammalian food on the ground. The prey is killed with a neck bite in the manner of the weasel. Birds often have their feathers sheared off near the base. Most of the birds caught by the pine marten are small species, such as tits and wrens, and the marten therefore avoids too much competition with the sparrowhawk, which prefers larger birds such as finches and thrushes.

The largest of the martens is the American fisher (*Martes pennanti*), which can reach 1m (39in) in length. Although it is a good climber like all the martens, it hunts mainly on the ground. Some fish may be caught, but the fisher actually feeds mainly on other mammals, including *M. americana*, beavers, and even porcupines. The latter are turned on to their backs by the fisher and attacked on the unprotected belly.

The kill
Left The stoat leaps on to the rabbit and goes immediately for the neck-bite (right).

Mongooses and Raccoons

The mongooses are small carnivorous mammals belonging to the family Viverridae. There are 36 living species, ranging from the size of a weasel to that of a large domestic cat. All are natives of Africa and Asia, with the Egyptian mongoose (*Herpestes ichneumon*) just reaching into southern Europe.

Mongooses are famed for their snake-killing activities, but in reality snakes play only a minor part in their lives. They are real opportunists, feeding on an extremely wide range of prey. The small Indian mongoose (*H. auropunctatus*) was introduced to the West Indies in the 1870s with the aim of controlling the rats in the sugar cane plantations. It did this to some extent, but it also found many of the native animals to its liking and a number of these became extinct soon after its introduction.

The adaptable mongooses
Mongooses can be found in all kinds of habitats, although most of them keep their feet firmly on the ground. The slender mongoose (*H. sanguineus*) is one of the few that climb trees with any regularity, while the water mongoose (*Atilax paludinosus*) is one of the few aquatic species, although all can swim when necessary.

Between them, the mongooses take all kinds of food, from worms and slugs, through numerous insects, especially grasshoppers, to snakes, birds, and rodents. The water mongoose takes large numbers of crustaceans and frogs, and is said to destroy many crocodile eggs, but we cannot accept the Roman idea that mongooses kill adult crocodiles by crawling down their throats while they are asleep and eating their way out through the stomach wall!

The smaller mongooses feed mainly on insects, with the abundant dung beetles providing easy meat in regions where large mammals, such as elephants and buffaloes, occur. Rodents provide most of the food supply for the larger species of mongooses.

The hunt
Many mongoose species hunt by day, using their sharp eyes and keen sense of smell. The insectivorous species search the ground methodically, turning over stones and logs and scratching vigorously in the ground when their noses indicate that something edible may be there. They pounce with amazing speed when they find an insect, and often shake it vigorously in the jaws before eating it. Mice are also snapped up by these primarily insectivorous mongooses, and killed with a well-aimed bite in the back of the neck. This is the method used by the more carnivorous species, and it suggests that the insectivorous species once fed mainly on vertebrates.

The more carnivorous species, typified by the slender mongoose, are extremely quick and agile, able to rush at and pounce on a rat without warning. Their sharp teeth produce almost instantaneous death as they are sunk into the victim's neck. Some species use fox-like cunning to lure victims out into the open and to mesmerise them before leaping onto them (see page 102).

Mongooses may hunt singly, in pairs, in small family groups, or even in packs. The banded mongoose (*Mungos mungo*), for example, may roam in packs of 50 or more individuals. Quite large prey, such as young antelope, may be killed by such a pack, but pack-hunting on this scale is much rarer than hunting in pairs. Two mongooses often co-operate to turn over stones in the search for insects or crabs; one moves the stone, while the other grabs the prey, and both then share the food. The Indian brown mongoose (*H. fuscus*) is believed to go fishing in pairs; one animal drives the fish shoals into the shallows, where another is waiting to pounce on them. A pair of mongooses may also work together to force a bird to leave its nest, thus allowing the predators to steal the eggs with relative ease.

To crack an egg
Eggs are greatly liked by mongooses, but the animals cannot always get their relatively small jaws around the eggs to break them. Instead, they throw the eggs against a hard surface to crack the shell. The banded mongoose, in common with a number of other species, picks the egg up in its front legs, raises its body on its back legs, and then throws the egg backwards, through the hind legs, to smash against a rock or similar object. Snails, crabs, nuts, and large pill millipedes may also be dealt with in the same way. The banded mongoose may also hurl stones at eggs that are too large and unwieldy to be picked up and thrown against a hard surface.

The water mongoose, however, has a different way of dealing with eggs and other hard objects, such as crabs. It holds the object in its front paws, rears up on its hind legs, and throws the object vertically down onto the ground.

Mongoose and cobra
Right Fur raised as a protection, a mongoose prepares to dart at a cobra which is already reared up ready to strike. Notice the fairly short fangs. The mongoose bites the snake just behind the head (centre picture) and holds on tightly. The snake thrashes about (far right), but it has lost.

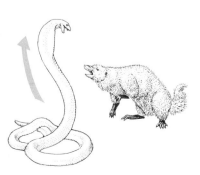

Banded mongoose
Left The banded mon-
goose (*Mungos
mungo*) may roam in
large packs, but it
more often hunts singly
or in pairs. Much of its
food consists of insects
and other small
creatures unearthed by
turning over stones.

Common raccoon
Below A common
raccoon enjoys a meal
of fish. Raccoons spend
a lot of their time in
and around the water
and eat quite a lot of
aquatic creatures. They
are sometimes said to
use their long,
sensitive fingers as
fishing lines, tickling
the water to attract
fishes, and then scoop-
ing them out with a
lightning movement.

To crack an egg
Left The Madagascan
narrow-striped mon-
goose (*Mungoictis
lineata*), not closely
related to the other
mongooses, lies on its
back and hurls the egg
with all four feet.
Left below The banded
mongoose raises itself
on its hind legs and
hurls the egg back-
wards through the legs.
This behaviour has
evolved from the
scraping for food.
Right The water
mongoose rears right
up on its hind legs and
throws the egg force-
fully onto the ground.

Plant food makes up the bulk of the diet of the common raccoon for much of the year, but animal food is dominant in spring and early summer. Voles and crayfish are the main prey, supple-mented by insects, snails, and earth-worms. The proportions taken vary from place to place, depending on the distance from water, but a certain amount of food is always taken in water. The crab-eating raccoon spends more of its time in the water, although it is by no means con-fined to water, and it takes a somewhat higher proportion of animal food than the common raccoon.

Both species are primarily nocturnal and, although their eyesight is reason-ably good, they find much of their food by touch — especially when they are feeding in the water. Like the water mongoose, raccoons have very long, sensitive fingers and they use these to find and catch their prey. Experimental work has shown that the front paws of the raccoon are so sensitive to tactile stimuli that the animals can distinguish between spheres of 2·5 and 2·64cm diameter (a difference of 0·05in) — a degree of sensitivity much the same as that of a human hand.

Captive raccoons very often develop the habit of dipping their food into water before eating it, and this led to their being described as fastidious because observers believed that they were actu-ally washing their food. Wild racoons do not wash their food, however, and observations by Malcolm Lyall-Watson at London Zoo have now shown that the actions of captive raccoons have nothing to do with washing the food. It is believed that dipping the food in water, and per-haps stirring it around for a while, fulfils the natural urge to catch prey in water. Dr R. F. Ewer also believes that dunking food and 'catching' it again may make it more acceptable to captive raccoons.

The mongoose and the snake

The snake-killing abilities of the mon-goose have been known and admired for centuries, but most of the accounts of fights between mongooses and snakes have been based on pairs of animals brought together in captivity. Mongooses certainly do eat snakes in the wild, as shown by the remains of snakes found in their stomachs and droppings, but the few observations that have been made in the wild suggest that mongooses ignore the larger snakes.

Nevertheless, the Indian mongoose (*H. edwardsi*) certainly can get the better of a cobra, largely by virtue of its in-credible speed and agility. The cobra is not a particularly fast striker; it raises its front end perhaps 50cm (20in) from the ground and then strikes at an object about that distance away. The attacking mongoose darts inside that distance, where it is relatively safe, and launches itself before the cobra strikes.

Although the cobra's venom is ex-tremely potent, its fangs are quite short and it must bite hard and long to inject an effective dose. It cannot do this to the active mongoose, which is further protected by its erected fur and a certain degree of immunity to cobra venom. The mongoose darts in again and again, and finally delivers the fatal bite to the much-weakened cobra's head.

A mongoose generally fares much less well if pitted against a viper, whose fangs are long and erectile and right at the front of the striking head.

The fastidious raccoon

Raccoons are truly omnivorous animals belonging to the family Procyonidae. There are two well-known species – the common raccoon (*Procyon lotor*) of North and Central America, and the crab-eating raccoon (*P. cancrivorus*) – and five lesser known species restricted to certain islands in Central America.

The Cunning Foxes

Foxes are medium-sized carnivores belonging to the dog family (Canidae). There are fourteen species, of which the best known is the red fox (*Vulpes vulpes*). This species ranges over Europe and North Africa, much of North America, and most of Asia. It is an extremely adaptable creature, able to live high in the mountains, among coastal sand dunes, in dense woodland, on the open fells and moors, and even in towns. It is equally adaptable in terms of diet and it will take virtually anything that is available to it.

David Macdonald, who has spent many years studying the fox's behaviour, found that the foxes in one part of Oxfordshire, in central England, were taking about half of their daily food requirements in the form of earthworms, with a fox eating as many as 146 worms in a single night. In the fells of Cumbria, in northern England, he found that the foxes ate no earthworms — no worm bristles were found in the foxes' droppings — and that they lived mainly on rabbits and grouse.

Earthworms and rabbits represent two extremes of the normal prey spectrum. In between there is a vast array of animal food available to and acceptable to the fox — slugs, beetles, grasshoppers, mice, voles, rats, and birds all figure prominently in analyses of the stomach contents and droppings. Foxes will also eat carrion and are partial to blackberries and other fruits.

Ample proof of the fox's ability to adapt its diet to the available sources came in the 1950s, when myxomatosis almost detroyed the rabbit population in Britain. The rabbit was known to be the major food of foxes in many parts of the country, and yet its virtual disappearance made little difference to the fox population. Analysis of fox droppings showed that the animal merely increased its intake of voles, birds and insects.

Although foxes will eat whatever they can get, they certainly do have food preferences when they are given the choice. Examination of stomach contents and droppings shows that the foxes eat far more voles (*Microtus* sp.) than mice (*Apodemus* sp.), even when the latter are more abundant in the environment. This could be, of course, because the voles are easier to catch — they cannot jump nearly as well as mice — but captive foxes eat dead voles much more readily than they eat dead mice, indicating that they really do prefer the voles.

A nocturnal prowler

The red fox hunts mainly at night, although there is much activity around dusk in the winter. It hunts alone, using its eyes, ears, and nose to detect prey

The fox's athletic mouse jump
Below The red fox detects a mouse or other small mammal in the grass and pin-points it position with ears and nose. A sudden leap high into the air, and the fox then lands right on top of the prey, holding it firmly to the ground with its front paws.

animals. It then stalks them carefully until it is close enough to pounce. Voles and mice, which tend to keep to concealed runways in the grass, are detected by sound and scent, and the fox pounces on them in a very characteristic manner. Having fixed their position, it leaps into the air and comes vertically down on them, with front paws and jaws together to pin the prey to the ground. This action has been called the 'mouse jump.'

Larger prey is generally caught with a simple snap of the jaws, and it may subsequently be thrown into the air and caught again or shaken vigorously. The fox does not seem to attack any definite part of the prey's body to kill it. Small rodents are swallowed whole without any 'preparation', but birds are roughly plucked by biting the feathers off near the base and leaving them lying about.

There are records of foxes charming or mesmerising rabbits by indulging in frantic bouts of leaping and dancing. The rabbits are 'intrigued' by the performance and they do not notice that the performing fox gradually gets nearer to them. Suddenly, the play stops and the fox grabs one of the rabbits. This is almost certainly something that some foxes learn to do. They are very playful animals. often frolicking in pairs or in larger groups even when they are fully grown, and it would not be difficult for the foxes to learn that such playful behaviour often attracts edible spectators. A performance might then be put on when a group of rabbits is seen.

Underground larders

The red fox often kills more prey than necessary to satisfy its immediate needs,

but this extra food is not usually wasted, for the fox generally buries it for another day. The fox carries the food to a suitable spot, and still holding the food in its jaws, digs a hole with its feet. The food is then pushed into the hole and the earth swept back in with the snout. The earth is also pressed down with the snout and the vegetation tidied up with a sweeping motion, resulting in a very well camouflaged larder.

David Macdonald has studied this food-caching behaviour with the aid of some tame foxes that were allowed to wander over an area similar to that patrolled by a wild fox. One vixen found 48 out of 50 mice that she had cached on previous occasions, but she did not find other food the experimenter had cached nearby and she was not very successful at finding her own caches when the

experimenter re-buried them one metre away. This suggests that the fox definitely remembers the spot where it has buried food and does not go sniffing around at random, although other experiments have suggested that the animal remembers only the general area in which a cache has been made.

Both sets of results suggest that a fox would not normally find the caches of another individual, although this does sometimes happen by accident. Macdonald tested this hypothesis with a pair of foxes and discovered that only the 'owner' – the one that buried the food – had any great success in finding it.

Macdonald's experiments also showed that his foxes could remember what they had cached in each hole. A hungry vixen taken into an area where she had buried both mice and voles went straight to the cached voles – proving that she had a remarkable memory as well as confirming the preference for voles over mice. In fact, the foxes were much more inclined to cache surplus voles than mice, often leaving the latter lying around.

Quite a lot of food is cached at dusk, when it is often very plentiful, and eaten later the same night. Other caches, however, may be left for days or even weeks. This is clearly of great value to the foxes in time of food shortage. The Arctic fox (*Alopex lagopus*), which inhabits regions where the weather makes hunting very difficult on many occasions, builds up very large food caches. One such cache was found to contain 36 little auks, four snow buntings, two young guillemots, and a large number of little auk eggs. Several caches of this size would certainly help the fox through the cold winter months.

The bat-eared fox
Most species of fox hunt and feed in much the same way as the red fox, although the actual prey species differ from place to place. The bat-eared fox of South and East Africa, however, is

unusual in that it feeds extensively on termites. It locates these in their underground tunnels with the aid of its huge ears, and then digs down to get at them. Beetle grubs are also avidly eaten, but these foxes do not rely entirely on insects, and Louis Leakey, who was better known for his discoveries of fossil men, recorded some fascinating examples of piracy against falcons. Groups of bat-eared foxes were seen watching falcons descend to the ground with pigeons, and the foxes attacked as soon as the birds landed. On each occasion the falcon tried to drive off one fox, but it had to release its prey to do so and another fox was able to snatch the prey and carry it off to the den, where all the foxes met up to enjoy the meal.

Such behaviour could easily be learned by foxes living in areas frequented by the falcons, but it is certainly unusual for foxes to cooperate in hunting activities. Recent work, again by David Macdonald, suggests that the red fox is a more sociable animal than we thought, but hunting is still very much an individual activity.

Where shall I put it?
Left A red fox looks for a suitable spot in which to bury a small mammal. The food may be dug up again and eaten within a few hours, or not for days.

The juggling fox
Above Mice and other small prey are often thrown into the air and caught again on several occasions before they are finally swallowed.

Jackals-The Hunting Scavengers

The jackals are close relatives of the dog and the wolf, and in many respects they look just like lean and undernourished wolves. There are three species, all averaging about 65cm (26in) in length, excluding the bushy tail, and all standing about 40cm (16in) high at the shoulder.

The golden jackal (*Canis aureus*) lives over much of southern Asia, in a small area of south eastern Europe, and over much of the northern half of Africa, extending down the east side as far as Tanzania. The black-backed jackal (*C. mesomelas*) lives throughout East and South Africa, and the side-striped jackal (*C. adustus*) can be found in most parts of Africa south of Senegal and Somalia. Most of our knowledge of jackal behaviour comes from studies of the golden and black-backed species in East Africa.

Not so cowardly
Like the hyena (see page 128), the jackal was once thought to be nothing more than a cowardly scavenger, clearing up the scraps from the lion's table. It certainly does scavenge – even to the extent of raiding village garbage bins – but it is not the coward that many people once thought. It fights off vultures and eagles, and even hyenas, and it will often dart in to snatch morsels from right under a lion's nose. Scavenging is clearly a part of the jackal's life, but its importance varies from place to place; in many areas the jackals exist mainly on food that they kill themselves, such as small antelopes, rodents, snakes, lizards, birds, and insects. Jackals are also fond of fruit, often raiding plantations for pineapples and other juicy titbits.

The golden jackal is mainly an animal of the open plains, while the black-backed jackal – also known as the silver-backed jackal – prefers the bush habitat. The two species overlap in the Serengeti region, however, and display some interesting differences in behaviour. On the open plains both species hunt Thomson's gazelles, which, together with the abundant dung beetles, make up the bulk of the diet. With so much live food available, scavenging is rare, and analysis of the jackals' droppings, or scats, has confirmed this; only about 3 percent of their volume represents large animals – food that could have been acquired only by scavenging.

In the marginal areas between the open plains and the denser bush, the golden jackal takes fewer gazelles and makes up for this by eating far more insects. The black-backed jackal, on the other hand, eats fewer insects in the marginal areas and turns more to rodents and other small mammals.

Small gazelles are less common in the

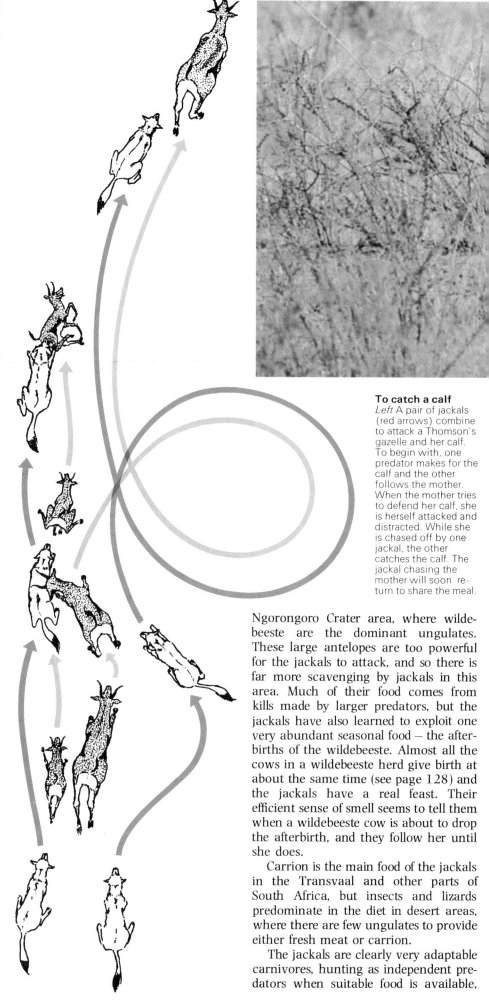

To catch a calf
Left A pair of jackals (red arrows) combine to attack a Thomson's gazelle and her calf. To begin with, one predator makes for the calf and the other follows the mother. When the mother tries to defend her calf, she is herself attacked and distracted. While she is chased off by one jackal, the other catches the calf. The jackal chasing the mother will soon return to share the meal.

Ngorongoro Crater area, where wildebeeste are the dominant ungulates. These large antelopes are too powerful for the jackals to attack, and so there is far more scavenging by jackals in this area. Much of their food comes from kills made by larger predators, but the jackals have also learned to exploit one very abundant seasonal food – the afterbirths of the wildebeeste. Almost all the cows in a wildebeeste herd give birth at about the same time (see page 128) and the jackals have a real feast. Their efficient sense of smell seems to tell them when a wildebeeste cow is about to drop the afterbirth, and they follow her until she does.

Carrion is the main food of the jackals in the Transvaal and other parts of South Africa, but insects and lizards predominate in the diet in desert areas, where there are few ungulates to provide either fresh meat or carrion.

The jackals are clearly very adaptable carnivores, hunting as independent predators when suitable food is available,

Scavenger, predator
Above A pair of black-backed jackals have found the remains of a springbok that had been killed by a cheetah. Such a carcase, probably detected by smell, would provide a substantial meal for the two jackals.
Right A black-backed jackal carries away a small mammal – probably to bury it for another day. Small mammals are heard among the grasses and pounced upon.

but equally happy to rely on leftovers under different conditions.

Jackal hunting strategies

Jackals can sometimes be found in quite large packs, but these are less common than they used to be and the animals are usually seen in small family groups. Most hunting is done by pairs or by individuals.

The black-backed jackal generally hunts in pairs throughout the year, but the golden jackal normally hunts in pairs only during the breeding season – between January and April in East Africa – when the parents work together to feed their offspring. Each pair hunts over a definite home range, although neighbouring families may share their hunting grounds and they may even join together to feed at a large carcase when food is scarce. Within the home range, however, there is a private territory where the young are reared, and this territory is strongly defended.

When hunting Thomson's gazelles or young impala – a common prey in South Africa – the jackals may stalk them carefully, but more often they use the typical dog technique of chasing the prey openly and wearing it down. They snap at its legs and eventually pull it down, but the two jackal species use different methods for despatching their victims. The black-backed jackal normally bites the throat, but the golden jackal goes straight for the belly and quickly eviscerates its prey.

Hunting in pairs is particularly efficient when the antelopes have young with them; one jackal attacks the mother and distracts her attention from her youngster, allowing the second jackal to take the youngster without hindrance.

The jackals eat their fill, and take enough for any cubs they have, and then bury the rest for another day. The food is usually buried over a wide area, one piece here and another piece there, in an attempt to foil other scavengers. But observations suggest that the jackals normally come back for it within about 24 hours. Food taken back for the cubs is swallowed and regurgitated later – a good defence against eagles, which might otherwise attempt to snatch food being carried back for the youngsters.

Smaller prey, such as rodents – which make up about 80 percent of the food of the golden jackal in India – are generally detected by the jackal's extremely good hearing, and they are often caught by the 'mouse jump' so familiar to fox-watchers (see page 102). The jackal leaps into the air and comes vertically down on the prey. The same method may be used to catch dung beetles and grass-hoppers, but beetles and moths may also be caught by leaping up at them in flight – showing that the jackals use their eyes as well as their ears in hunting.

The sense of smell is also important, especially in the detection of carrion at night. Jackals may find quite a lot of carrion in the daytime by watching the circling vultures and following them when they drop down to the ground, but most carrion is eaten at night.

The Omnivorous Bears

Apart from the carnivorous, predatory polar bear, the bears are a congenial group of ambling foragers that feed on both plants and small animals. They are the largest carnivores alive, shuffling along on the flat soles of their feet on the lookout for any opportunity that presents itself. Despite their bumbling appearance, they can be surprisingly agile and fast when the need arises, such as the capture of a startled fawn or bird. Although their hearing is reasonable, they tend to be short sighted and rely on their keen sense of smell for the location of food.

Being generalized omnivorous feeders, their teeth are unspecialized. The cheek teeth are broad and flat, enabling them to grind up hard vegetable matter and animal bones, but they have strong canines for fighting and gripping struggling prey. Each foot has five powerful claws and these are especially big and strong on the forepaws for tearing at foliage or bee nests, or for digging open cement-like termite nests. The shaggy coat provides them with insulation and protection from many of the stings of angry defending bees and wasps. Bears consider a few stings a low price to pay for fat juicy grubs and sweet honey. They can stand upright on their hind legs to reach any delicacy high in the trees, the grizzlies towering up an impressive 3m (10ft), and many of the smaller bears are excellent tree climbers, digging into the bark with recurved claws.

All bears are solitary creatures, wandering alone except during the mating season or when the cubs accompany their mother. Most species live in forested mountainous regions, but bears are found in habitats ranging from the north pole to the tropics and occur in the Americas, Europe, and Asia. They are most numerous in the northern hemisphere, only a few species living in the southern hemisphere. Those in the northern temperate regions often sleep in their dens for long periods in winter, relying on their thick layer of body fat as a food store until the next spring.

Fishing grizzlies

The huge grizzly bears are usually solitary, wandering about their vast home ranges some 72km (45 miles) in diameter, but when the sockeye salmon start to travel upstream to spawn the irascible grizzlies congregate in large numbers along favoured river banks. Each bear stakes out its fishing ground, which is ferociously defended so that the larger and more powerful bears get the best spots. These are usually in the shallows where the floundering salmon are easily trapped and flipped out of the water onto the river bank with a swipe of the paw; or part way up a waterfall where the salmon land after leaping the lower section and then struggle desperately against the current to gain the top of the fall. At this point the current keeps the exhausted fish almost stationary so it is easily plucked from the water. The grizzly usually does this by scooping it out with a forepaw and grabbing it quickly in its jaws, impaling the slippery fish on its piercing canines so it does not wriggle free. Sometimes the bear will plunge its muzzle directly into the foamy water and snatch out the fish with its jaws. The captive is carried to the bank to be devoured with relish. During this season the bears dine almost exclusively on salmon, taking full advantage of an easily available source of protein.

Grizzlies are one of a group of bears known as brown bears (*Ursus arctos*). The Alaskan, or Kodiak, bear is another race of brown bear even larger than the grizzly and can weigh up to 780kg (1716lb). Smaller brown bears are also found in northern Europe and parts of Asia. All are partial to salmon in season but feed mainly on plant leaves and fruits, taking insects (including hibernating ladybirds, caterpillars and caddis fly

larvae), rodents, birds, squirrels and deer when available. Even bison have been killed by large grizzlies.

Park scavengers

The American black bear, *Ursus americanus*, is well known as a scavenger overturning litter bins in National Park campsites. The colour of these bears can range from black through chocolate to

The fishing bears
Top and left Brown bears are extremely fond of fish and frequently flip salmon and other fishes from the streams with their paws. The fishes may be flipped straight on to the bank, where they are stranded, or they may be flipped straight into the mouth. The bears often congregate in large numbers around the shallow spawning streams of the salmon, and each bear defends a particular stretch of the bank. The bears in the top picture are Kodiak bears, from Kodiak Island in Alaska. They are the largest race of brown bears.

cinnamon and even, occasionally, greyish-white. Black bears are smaller than brown bears, with less shaggy coats, shorter claws and smaller hind feet. Being excellent tree climbers, they shin up and down tall tree trunks in search of succulent leaves and berries or fat grubs and sweet honey. They often use their strong claws to tear off pieces of bark to expose the fat grubs in the channels underneath. These are licked up with gusto.

Bear or yeti
The Asiatic black bear, *Selenarctos thibetanus*, of the mountainous regions of Asia is about the same size as the American black bear but is reputedly more aggressive, raiding livestock and killing ponies. It may be the footprints in the snow of a large Asiatic black bear that have given rise to the speculations on the existence of the Himalayan Yeti whose hunting habits are still unknown!

The sucking sloth bear
The sloth bear, *Melursus ursinus*, lives in the forested areas of India and Ceylon, its range only just extending into the Hima-

The American black bear with a kill

Above The American black bear is well-known to visitors to North America's National Parks, where it frequently scavenges around campsites.

Although it gets a lot of food by scavenging, it does not hesitate to kill when the opportunity arises. This bear has killed an unwary beaver.

layan foothills. It is almost entirely insectivorous and gathers up its tiny prey by giving a remarkable imitation of a vacuum-cleaner. After using its long claws to dig out a termite nest, the sloth bear closes its nostrils, purses its long mobile, naked, protrusible lips and blows through a gap in its teeth where the upper

pair of inner incisors are missing. This blows away unwanted dust and extraneous debris and the termites are then sucked up through the same tube.

The lazy sun bear
The smallest bear is the Malayan sun bear, *Helarctos malayanus*, to be found sunbathing and sleeping in a tree nest during the day in the forests of Southeast Asia. Living in a warm tropical or subtropical region, it has a short, glossy coat with a whitish or orange pattern on the chest. The soles of its feet are naked, giving it a better grip as it clings to the tree bark with its long curved claws. The bear has a peculiar gait when on the ground, because its legs are turned inwards to avoid blunting and damaging the claws. Like all bears, it is an omnivorous feeder but it finds termites a great delicacy. The sun bear is not a greedy vacuum-sucker like the sloth bear, instead it breaks open the termite nest and then daintily dips its paws into the seething mass of insect life, allowing the termites to swarm over them. Lifting a paw, the bear licks off the insects and plunges the paw back for another helping.

The Solitary Tiger

The tiger vies with the lion for the title of the world's largest cat. Both species can reach lengths of about 3m (10ft), including the tail, and some tigers perhaps reach overall lengths of about 4m (13ft). Tigers are found mainly in eastern and southern Asia, where they inhabit a broad band sweeping down from the Sea of Okhotsk, through eastern China, and on to Indonesia and India. Small populations also occur around the Caspian Sea and on the Turanian Plain of southern Russia. Eight subspecies are recognized, but the animal is extinct in many parts of its range and not common anywhere today.

Tigers inhabit virtually every kind of habitat within their range – deciduous and evergreen forests, grass swamps, bamboo thickets and other scrub, and savanna – and their prey is likewise very diverse. Wild pigs and various kinds of deer constitute their main food nearly everywhere, and tigers are so fond of wild pigs that they even dig frozen carcases from the snow in Siberia. Captive tigers have shown that they do really prefer pig meat to other flesh, thus suggesting that it is not just the ease of capture that leads them to take so many pigs in the wild.

Among the various kinds of deer eaten by tigers are the red deer, the chital or spotted deer, and the muntjac. In Indochina the sambar deer is a favourite prey, although it is a very alert one and liable to warn other potential prey with its loud, ringing bark. The sambar leaves the forests to browse and graze in clearings at night, and it is then quite an easy target for the tiger. Many antelopes are eaten, including the large nilghai. Abandoned cultivated areas are favourite haunts of this species, and the tiger is also fond of such places, especially if there are ruined buildings in which it can shelter by day.

Other commonly eaten prey include domestic cattle and buffalo, horses, monkeys, and even porcupines. The tiger will also kill and eat other carnivores. It has a strong liking for the smaller cats, including domestic ones, but it has been known to attack bears when the opportunity arises.

The lone prowler

Like most of the other cats, the tiger is a solitary prowler. A male has a territory covering as much as 5000km² (1950mi²) and it patrols the area regularly, perhaps taking two weeks or more to complete a tour. Tigresses have smaller home ranges, which generally overlap and which are not defended. Several tigresses may live within the territory of one male, and all mark their boundaries with urine. The animals rarely meet, although

two or three sometimes gather when one of them has made a particularly large kill.

Tigers are active by day and night, but in the warmer parts of their range they do most of their hunting after sunset; both predator and prey lie up in the shade during the heat of the day. Darkness gives the tiger extra cover, not that it is poorly camouflaged in daylight, and there is the added advantage that most of the birds and monkeys are asleep and less ready to sound the alarm. Even so, the forests are regularly woken by the scream of a peafowl alerted by a prowling tiger, and this is quite enough to send the deer fleeing for their lives.

The hunt

Whether hunting or merely patrolling its area, the tiger usually keeps to well-marked pathways through the forest or grassland. It sometimes ambushes prey on their regular pathways, especially near water holes, but it generally trots quietly along until it detects something interesting. A great deal has been written, and many conflicting opinions have been expressed, concerning the relative importance of the tiger's three main senses – sight, hearing, and smell. All three are certainly used in hunting, and

their value probably varies with the type of terrain, but it is clear that sight and hearing are the major hunting senses.

The importance of the eye can be appreciated from its large size and the presence of a tapetum behind the retina (see page 69). When the animals are picked out in a searchlight at night, the tapetum reflects the light and makes the eyes shine with a beautiful greenish gold colour.

Having detected potential prey, the tiger begins its stalk, which is designed to bring it to within striking distance. Dropping down with its belly almost brushing the ground, the tiger seems to flow along just like the much smaller leopard. The head is also held very low, with the jaws open, and 60cm (24in) of grass is sufficient to hide the animal from its prey. The tiger's stripes help to camouflage it even among scattered tufts of grass. Like all stalking cats, the tiger is very careful about where it puts its feet, but its great cushioned pads allow it to approach its prey in virtual silence on any surface – even on dead leaves.

When stalking deer, the tiger must get within about 25m (82ft) of them – and preferably nearer – without being noticed. Deer often feed out in the open, and if the cover stops more than about

The solitary stalker
Left The tiger's beautiful striped body merges into the dappled undergrowth as the animal silently stalks its prey through the forest. It kills mainly antelope, deer, and pig, but will take birds, rodents, or even other cats.

The kill
Right The stealthy stalk of this forest-dwelling tigress has brought her within ambushing range of her monkey victim. After a pause, she launches herself at the prey, knocking it to the ground and killing it with one bite.

Territorial marking
Right Having made a kill, tigers often scent-mark their own territory by spraying urine over a tree stump, a mound or a rock. This marking warns other tigers that these hunting grounds are occupied.

Tiger's teeth
Below The casual yawn of this predator exposes its massive canine teeth which grip and kill the prey. Large prey may be suffocated by a throat grip, but the teeth often penetrate the skull or the spinal cord of smaller victims.

25m from them the tiger often gives up the hunt, as if knowing that success is unlikely. When sufficient cover is available, the tiger makes the most of it and gets as close to the deer as possible. It then stands, often with one front paw lifted and the tail held straight out behind, and sways gently to achieve perfect balance before the final rush. With immense bounds the tiger surges forward onto its victim. In some instances, a single bound is enough to bring the tiger into contact with its prey.

The kill
The tiger almost always attacks from the rear and, when attacking hoofed prey, it goes for the neck or throat. The final assault is sometimes described as a leap, but it is really more of a lunge, for, as with nearly all cats, the hind legs do not normally leave the ground (see page 114). One front leg is generally thrown around the shoulders of the prey, and the tiger then bites deeply into the neck or throat. This action brings the prey down headfirst, and often breaks its neck at the same time.

The neck is certainly twisted as the prey falls, and if it is not broken the animal soon dies through strangulation as the tiger's teeth bite deeply into the throat. The tiger ends up kneeling down by the prey's head, well away from any flailing hooves and also on the opposite side to any horns or antlers. Smaller prey is much less of a problem, and is merely pounced upon in typical cat-like fashion.

Prey killed in the open is generally taken to a secluded spot before the tiger starts to eat. Such is the strength of the tiger that it has no trouble in dragging the carcases over distances of 500m (1640ft) or more — trailing them either beside or underneath its own body. One tiger was seen to do a high jump of over 4m (13ft) while carrying a carcase of about 70kg (154lb) in its jaws.

Like many other carnivores, the tiger begins its meal with the rump steak as a rule, tearing chunks off with its canines and using its razor-sharp carnassial teeth (see page 75) to slice the meat into pieces that can be swallowed. Vast amounts can be eaten at one sitting; one large Manchurian tiger was reported to have eaten about 50kg (110lb) in one session, although 25kg (55lb) is probably a more normal amount.

Several meals may be taken during the night, and if anything remains in the morning the tiger may bury it or even dump it in a pool. The animal then lies up somewhere in the vicinity to guard its larder, but it rarely revisits the cache until the next night. Virtually everything is eaten in the end, including most of the bones. Only the skull and the lower parts of the legs remain to show where an ungulate has been killed.

A large kill, such as a sambar deer, may last a tiger three nights. By the third night the meat may be very 'high', but this seems only to add to the tiger's enjoyment of it, for the animals readily eat maggot-infested carrion that they find during their travels. One large kill probably provides enough food for one week, with the tiger refraining from deliberate hunting for several days after finishing the carcase.

An intelligent hunter
There is no doubt that the tiger learns much of its hunting technique while growing up with its mother and brothers and sisters, and that it continues to learn throughout its adult life. For example, it very soon learns to abandon a stalk if a deer detects it too early and gives the alarm call; the tiger is very fast over a short distance, but it cannot give much start to a deer if it is to have much chance of overtaking it.

Some observers have suggested that the tiger even learns to mimic the mating calls of deer and thus lure the deer towards it, but this seems unlikely. It is probable that these observers have misinterpreted the roar of excitement that the tiger often emits on beginning its final assault on its prey.

One macabre, but nevertheless interesting piece of learned behaviour was discovered during the Vietnam war: tigers learned to associate gunfire with the presence of food in the form of dead bodies and regularly appeared after skirmishes. This leads us on to the topic of man-eating, a habit which occurs spasmodically in most tiger populations. Most man-eating tigers are old or injured individuals that find it difficult to catch game or even cattle, but the habit may also come about through a chance encounter with a corpse if the tiger finds the human flesh to its liking.

The Cheetah-The Feline Greyhound

The cheetah is well known as the fastest animal on four legs, although this oft-quoted statement needs some qualification: the animal is certainly the quickest off the mark, reaching about 70kph (43mph) in just two seconds, and with a top speed of over 110kph (68mph) it is the fastest runner over a short distance, but the cheetah does not have the stamina of many other animals and it cannot maintain high speeds for more than about 400m (1300ft). It is built entirely for sprinting, with no surplus weight anywhere. Its head is the smallest of any cat in relation to the rest of its body, its legs are long and slender, and its amazingly flexible backbone arches up as the hind legs come forward and gives extra length to each stride.

High-speed hunter

The cheetah's hunting methods are very different from those employed by the lion (page 126) and the leopard. The lion and the leopard prefer the more wooded areas of the savanna, although they do venture onto the open plains from time to time, but the cheetah is purely a hunter of the plains. It has little chance of stalking up on its prey unseen, and it must rely entirely on its tremendous acceleration and speed.

The cheetah also differs in doing almost all of its hunting during daylight hours – a direct result of its high-speed methods, which are simply not suitable for use in the dark. Most of the hunting is carried out early in the morning or late in the afternoon when temperatures are not too high, because heat, as we shall see, is a constant problem for the cheetah.

Cheetahs will eat a wide variety of animals if they get the chance, but in practice their menu is rather restricted. Hares and impala are regularly taken in some places, and a pair of cheetahs working together will even bring down a young wildebeeste, but the great bulk of the cheetah's food consists of Thomson's gazelles. An adult gazelle weighs only about 20kg (44lb) – less than half the weight of an adult cheetah – and it is therefore quite easy for a cheetah to bring one down. Thomson's gazelle itself has a good turn of speed, however, and actually catching one is not easy.

Approaching the prey

Cheetahs have good eyesight and they can spot their prey from considerable distances. They often climb onto termite mounds to get a better view of their surroundings. When a gazelle is spotted, the cheetah ambles towards it quite openly for a while. Its head may be lowered, but its legs remain straight and its whole attitude is markedly different from that of the stalking leopard.

If the gazelle looks up, the cheetah 'freezes' and, as long as it is not too close to the gazelle, the latter seems to take little notice. The critical distance is about 50m (164ft), and the cheetah must get within this distance without disturbing its prey if it is to have any real chance of catching it. At about 50m the cheetah breaks into a trot and quickly speeds up to a full sprint.

The gazelle is on the move by this time,

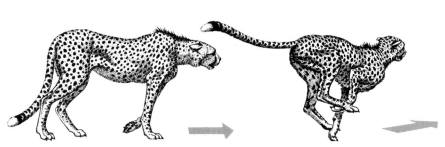

A sharp-eyed hunter on the lookout
Left The cheetah relies on its keen eyesight to scan the surrounding plains for a Thomson's gazelle that could provide it with a meal.

Caring for future hunters
Above After mating, the solitary male cheetah leaves the female to bring up the cubs. It is she who teaches them to hunt.

running in a straight line to start with but jinking from side to side as the cheetah closes in on it. This agile jinking may cause the fast-moving cheetah to overshoot its target, but it can also work in the cheetah's favour, allowing it to cut a corner and get even closer to the prey.

The kill

When the cheetah does catch up with the gazelle it brings it down by catching hold of the hind legs or by delivering a powerful blow to the rump. It then quickly grabs the victim's throat and strangles it, but it does not usually begin to eat right away. It exerts so much energy during the chase that it has to rest for anything up to 15 minutes, panting hard to get its breath back and to cool itself down. Experiments have indicated that the cheetah's body temperature rises to the dangerously high level of 40.5 °C (105 °F) during a 400m sprint. Any further exertion would damage the brain, and the cheetah must give up the chase.

The gazelles do not have this overheating problem, thanks to a special network of blood vessels near the base of the brain. Blood which has been cooled in the nasal airways flows close to the brain-bound blood, cooling it sufficiently to avoid any damage to the brain. The gazelles can thus go on running, and they are quite safe if they are not caught within the first 400m. A 50m start is usually sufficient to ensure the escape of an adult, although young gazelles are not so quick and they are much easier prey for the cheetah. Thomson's gazelles comprise 90 percent of the prey taken by cheetahs.

Observation of the cheetah's hunting activity suggests that about half of the chases are successful. This is a far higher proportion than among the lions and leopards, but Brian Bertram, who has studied all three cats in Africa, suggests that this may indicate less optimism on the part of the cheetah rather than greater efficiency. So much energy is expended on the chase that the cheetah needs to be more certain of success than the lion or leopard, which stalk 'lazily' up to their prey.

When a kill has been achieved and the cheetah has regained its breath, it tends to bolt down its food. It cannot enjoy the luxury of a leisurely meal, for there are always hyenas and often lions waiting to move in and, being lightly built, the cheetah is unable to defend its prize. Even vultures will drive a cheetah from its dinner on occasion. In this respect, it is interesting to notice that the cheetah generally starts with the protein-rich rump steak — taking the best part first, almost as if it expects to be robbed of the rest. The other cats tend to rip out the entrails and eat those first.

Even if it is not disturbed, the cheetah tends to waste quite a lot of its food, especially if it brings down a large animal. It eats its fill, but displays none of the leopard-like behaviour of keeping the rest for later. The food is left for the jackals and other scavengers.

Family life

A solitary cheetah probably kills every two or three days, but a female with young cubs has to kill every day. The cubs — usually two or three, but sometimes four or five — stay with their mother for about 18 months, during which time they learn all the tricks of the hunting profession. The females usually adopt solitary lives soon after leaving the mother, but brothers often stay together for much longer and sometimes form permanent partnerships. A stranger may also join up with such a group, forming an efficient hunting team of three or even four individuals that hunt over large areas of the plains. By working as a team, the cheetahs can bring down larger prey, but they still rely mainly on gazelles and follow the herds wherever they roam.

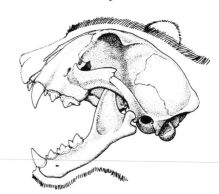

The bounding killer
Below The long legs and flexible spine make the cheetah the fastest sprinter of the plains. Gaining on its quarry, the cheetah trips it with a side-swipe of its paw. The luckless victim is then strangled by a throat grip as the long canine teeth puncture the tissues.

Shearing teeth
Above For lightness the cheetah has a small, short skull equipped with long pointed canines to grip its struggling prey and a few razor-sharp cheek teeth, the carnassials. These shear through the soft meat on the carcase, cutting it into lumps to be swallowed.

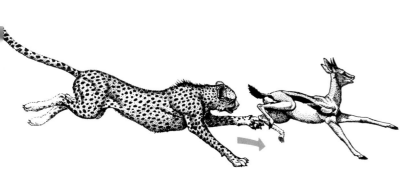

The Leopard – The Lonely Hunter

The leopard (*Panthera pardus*) has a much wider distribution than any of the other large cats. It ranges throughout most of Africa and virtually the whole of southern Asia, including much of eastern China. It is also much more adaptable than most of the other cats, and it occupies a far greater variety of habitats. In some parts of Africa it shares the grassy plains with the lion (see page 124) and the two species compete to some extent for food, but the leopard's hunting methods mean that it needs more cover than the lion. It prefers the more densely wooded parts of the savanna, but it can also live in the semi-desert as long as there are plenty of rocks to provide shelter. It also ranges high into the mountains. The leopard is most at home and most abundant, however, in the dense forests of the tropical regions.

Except when a female is rearing her cubs – three may be born, but only one or two are normally reared in the wild – the leopard is a truly solitary animal with a distinct territory. The territorial boundaries are defended against members of the same sex, but male territories often overlap with or even include several female territories, thus allowing the animals to meet for mating.

The opportunist

When it comes to finding food, the leopard is a real opportunist, with a very long list of prey species, even in a small area. On the African savanna, its food ranges from the innumerable dung beetles, through birds and rodents, to the various kinds of antelopes and even young zebras. Bush-pigs are also relished, and baboons are taken when the opportunity arises, but the efficient social organization of baboons and the powerful teeth of the adult males make it unlikely that leopards ever get very fat on baboon meat. Dogs are taken quite frequently in India, as are jackals, mongooses, and some of the smaller cats.

The leopard is unusual in taking so many other carnivores as part of its diet, but its ability to exist on such a wide variety of prey is clearly an important factor in its success as a species. Carrion is readily eaten when discovered.

The stalk

When a leopard sees a potential victim, its normal reaction is to conceal itself to the best of its ability. This usually involves flexing the legs and dropping the body down very close to the ground, and making use of whatever cover is provided by surrounding rocks or vegetation. The spotted coat of the animal gives it additional camouflage, and the leopard begins to move stealthily towards its prey, gliding almost snake-like from one

piece of cover to the next. Its eyes remain firmly fixed on the prey and if the latter shows any sign of anxiety the leopard freezes in its tracks until the prey either bounds away or relaxes again. The leopard is not a sprinter like the cheetah, and it must get really close to its prey before making its final leap or dash to make the kill.

Many stalks end in failure because the prey detects the approaching leopard before it is within pouncing range, and quite often this happens because the leopard appears to pay no attention to wind direction; although its visual camouflage may be perfect, the leopard may approach from upwind of the prey and thus give an antelope warning.

When a leopard does get within striking distance, it either rushes at the prey or leaps onto it with a mighty bound. The outstretched legs and extended claws bring the victim to the ground and the leopard quickly administers a fatal bite with its powerful teeth.

Hunting and stalking may be carried out at any time of day or night, although most hunts take place around dusk and dawn, and provide the leopard with the bulk of its food, but the leopard does lie in ambush for prey on occasion. It frequently lies up in vegetation close to well-used animal tracks and pounces on passing animals. The leopard is also said to ambush prey by lying on horizontal branches and dropping onto animals

The jungle jaguar
Below The beautiful, rosetted jaguar usually lurks unseen among the dark, dappled foliage, never far from streams. Its usual prey are birds, forest deer, pigs and rodents, but it also wades deep into swirling waters to fish and to bathe.

Larder in the trees
Right The stealthy leopard kills mainly at dusk and dawn and, having eaten its fill, drags the carcase up into a tree out of reach of marauding ground scavengers. Here the cat can rest in the shade, guarding its kill against vultures.

passing underneath, but this is not common behaviour, even though the leopard is undoubtedly a fine climber.

Larder in the trees

Small prey, which is generally caught by young or infirm leopards, is usually eaten on the spot, but larger prey, such as gazelles and other antelopes, is usually taken to cover to be enjoyed at leisure. In Africa the prey is very often carried up into a tree, where it is safe from other predators and scavengers, notably the spotted hyenas (see page 128). This species is absent from Asia, and the leopards there are much less inclined to carry their booty into the trees. The ease with which a leopard can carry an antelope weighing almost as much as itself high into a tree shows just how strong this cat is. The heavy muscula-

ture of the neck is undoubtedly related to the habit of carrying prey.

An impala or other medium-sized antelope will provide a solitary leopard with food for three or four days, after which just the skin and bones can be seen hanging from the tree larder. Unlike most other cats, the female leopard does not take her growing cubs on hunting expeditions with her; she catches the prey and stores it, and then fetches the cubs to share the meal. Leopard cubs thus have to learn to hunt alone, and they start with the very small prey and gradually work up the scale as they get larger and stronger.

The jaguar

The jaguar is the South American equivalent of the leopard, although it is a rather more heavily-built animal -

weighing up to about 150kg (330lb) compared with the leopard's 80kg (176lb). Females are not much more than half of these weights. The jaguar's coat pattern is also distinct in that the rosettes of spots nearly all contain a central dark spot.

Like the leopard, the jaguar inhabits a variety of places, but it is most common in the dense forests. It spends a lot of time in the trees, where it catches birds and small mammals, but the bulk of its food consists of deer, peccaries, and the giant, guinea-pig-like rodents called capybaras. Jaguars also enjoy splashing and bathing in the water and they catch a considerable number of fishes. Some reports suggest that they 'deliberately' flick the tail in the water to attract fishes, and then scoop them up to the mouth with their huge paws.

Some Lesser Cats

There are 37 accepted species of cats in the world, all belonging to the family Felidae. Five of these are generally known as the big cats. These are the lion, tiger, leopard, jaguar, and snow leopard. The rest are collectively known as the lesser cats, although some, such as the puma, or mountain lion, may be as large as the leopard. The essential difference between the two groups lies in the larynx — not an earth-shattering difference, perhaps, but one that explains why the big cats can roar and the lesser cats cannot.

Woodland stalkers

With the exception of the cheetah (see page 110), the lesser cats are a very uniform group, although their sizes range from the puma, of which an average male weighs about 90kg (198lb), to the black-footed cat (*Felis nigripes*), which averages less than 3kg (7lb). Lesser cats can be found in almost every habitat outside the polar regions, but the majority of the species are woodland creatures. They feed primarily on birds, rodents, and rabbits. Most can climb well, and many hunt regularly in the trees. All have an extraordinarily good sense of balance.

As can be gathered from the large eyes, backed by a well-developed tapetum (see page 69), the cats are mainly nocturnal, with peaks of activity just after dusk and just before dawn. The animals also depend a great deal on their acute hearing. Compared with that of dogs, the cats' sense of smell is poorly developed — the short snout of the cat bears witness to this — and the animals rarely follow scent trails when hunting. The whiskers, or vibrissae, help the cats to feel their way around at night.

When prey has been detected, the typical cat begins to stalk forward with the slinking attitude already described for the leopard and the tiger. It may stop once or twice, remaining absolutely quiet and concentrating hard on the position of the prey. When in position for the final assault, the cat stops again and watches the prey intently, with tail twitching and feet itching in anticipation of the kill — a sight familiar to anyone with a cat in the household. The cat then darts forward and lunges at the prey, but keeps both hind feet on the ground so that it can compensate easily for any slight change in the position of the prey during the lunge.

Small prey, such as mice, are caught in the jaws alone, but larger victims may be pinned down by the front feet as well. The needle-like claws are ideal weapons for this job. The cat's sharp canine teeth administer a deep bite in the back of the neck, and the prey usually dies very quickly as its spinal 'cord is ruptured.

If it continues to struggle, the cat may release it and then take a fresh hold, but those species that feed largely on birds or fish do not normally release their prey, however much it struggles: to do so would mean to lose it.

The deep neck bite seems to be a basic habit of all cats, but the big cats and a few others have abandoned it in favour of the throat bite when dealing with large ungulates. This enables the cats to avoid injury from the prey's horns or antlers (see page 109). Smaller victims are dealt with by the neck bite.

Lesser cats eat their victims in much the same way as the big cats, but their front teeth may not be strong enough to tear out chunks of flesh and the smaller species tend to attack their food with the side of the mouth, using the carnassial teeth to carve the meat. Even domestic cats do this, although their tinned food needs neither tearing nor slicing. Mice and other small prey are eaten completely, although the stomach may be discarded, but the bones of larger victims are left after being cleaned by the incisor teeth and the rough tongue.

Some specialists

The serval (*Leptailurus serval*) is a beautiful African cat with very long legs, but it does not use its legs for speed as most other long-legged animals do. The paws are very mobile, and the front legs are clearly adapted for probing into burrows and crevices to catch small rodents. A serval often waits by a burrow until it hears the rodent coming up, and then a lightning strike with a paw hooks the

Spotted roost raider
Left The ocelot is an agile tree-climber, often taking roosting birds under cover of darkness. Even small birds are plucked before they are eaten.

The alert hunter
Below The serval relies on its acute sense of hearing to locate mice in their burrows. Its long limbs and mobile paws are used more for dragging prey from their retreats than for running.

Expert bird-catcher
Below The caracal is particularly adept at rearing up and snatching birds from the air. With one movement, the bird is clasped in the paws and drawn down to the jaws.

Bobcat the fisherman
Above The bobcat (*Lynx rufus*) eats rabbits and rodents, but is not averse to hooking trout from the water.

prey out and often flings it into the air. The serval pounces on it immediately, often before it has landed.

This technique is clearly very useful in the serval's savanna home, where small rodents disappear into the long grass as soon as they leave their burrows, but the serval does not depend entirely on this method. It stalks much of its prey through the grass in typical cat fashion, using its large ears as much as its eyes to guide it. Its long legs also enable it to jump well, and it frequently searches areas of long grass with a series of high, arching leaps. Prey is more easily seen from above, and the cat pounces on it with the next leap.

Its leaping ability also enables the serval to snatch low-flying birds from the air, and in this respect it resembles the caracal, or desert lynx (*Caracal caracal*). Both cats leap up with the body almost vertical and either knock the bird to the ground or catch it in a paw and sweep it to the mouth. Many of the birds caught are those that are disturbed on the ground, and the caracal is so fast that one is known to have caught ten pigeons in one second when it encountered a flock on the ground.

The margay (*Leopardus wiedi*) inhabits the forests of Central and South America and it is the most arboreal of all the cats. It even climbs down tree trunks head first like a squirrel. Most of its hunting is done in the trees, stalking along the branches to pounce on unsuspecting birds and other animals. The margay may also drop onto prey on a lower branch. Its large eyes are often thought to be an indication of nocturnal activity, but the dense forests are always rather dim and the animal may not be markedly nocturnal.

The ocelot (*Leopardus pardalis*) is a larger version of the margay living in wooded areas of tropical and subtropical America. It hunts by day or night, generally taking birds or small mammals on the ground. It stalks its prey until within striking range, but it does not

stop every now and then as most small cats do; birds tend to spend only short periods on the ground, and a fast stalk is much more likely to be successful than a slow one.

Ocelots also differ from other small cats in the way that they pluck the feathers from their avian prey. Most cats merely pull out a few of the larger feathers and eat the rest, but the ocelot goes in for careful and almost complete plucking, indicating a much higher level of reliance on bird prey than is found among other cats. The margay may also pluck its avian prey, but little is known of its feeding habits.

Pallas's cat (*Felis manul*) lives on the cold Steppes and mountains of Central Asia and it has all the characteristics of mammals living in cold climates – a thick-set body clothed with dense fur, and short appendages to reduce heat loss to a minimum. Perhaps its most interesting adaptation, however, is the very flat top to its head. The eyes are almost at the top, and the ears are well out to the sides. This unusual design is an adaptation of the cat's open habitat, in which the only cover is provided by rocks and scattered tussocks of grass or low shrubs. The animal hunts by sight, usually at night, and it can peer over the cover without showing much of its head and scaring away the hares and pikas on which it feeds.

As a final example of the more specialized lesser cats, we can look at the fishing cat (*Prionailurus viverrinus*) of Southeast Asia. Most cats will eat fish when given the chance – fish used to be the staple diet of the domestic cat at one time, and several species have been seen to flip fishes from the water with their paws (see page 112) – but few small cats are willing to go right into the water for their food. The fishing cat is an exception. It inhabits marshes, streams, mangrove swamps, and similar places and the scattered information available suggests that, as well as fishing expertly with its paws, it regularly plunges its head into the water to snatch food. It may even dive after prey, although it certainly does not confine itself to fishes. Water snails, crustaceans, frogs, birds, and small mammals are all caught in and around the water and the fishing cat has not lost any of its stalking abilities through its inclination to fish for its food.

Rock prowler
Left Pallas's cat lives in the cold, rocky areas of central Asia. Long hair on the lower part of the body keeps out the cold, while its very flat head allows it to peer over rocks to search for prey without being noticed.

Australia's Pouched Killers

Apart from the bats and various small rodents that must have 'hitch-hiked' from Asia on logs and other floating debris, Australia's only native mammals are the egg-laying monotremes and the marsupials. The monotremes are represented by the spiny anteater, or echidna, and the duck-billed platypus, while the marsupials, or pouched mammals, are a much larger group with about 170 species in Australia and New Guinea and about 60 species in the Americas.

Adaptive radiation

With no competition from the more advanced placental mammals in other parts of the world, Australia's marsupials were able to spread into all available habitats and adopt most possible feeding habits. There are herbivores, carnivores and insectivores, just as there are elsewhere. Biologists call such a development an adaptive radiation, for the animals radiate from a basic ancestral form and become adapted to several different ways of life. The same opportunities were available to the marsupials as were open to placental mammals elsewhere, and we find that many of the placental mammals have direct marsupial equivalents. There are monkey-like phalangers in Australia's trees, for example, and mole-like marsupials under the ground. There are also marsupial anteaters, while the kangaroos are the equivalents of the antelopes and other grazing animals. The kangaroo body is very different in shape from that of the placental grazer, but it meets the same requirements of speedy movement over the ground. The head and jaws of the kangaroo show marked similarities with those of the placental grazers because both feed in a similar way.

The Tasmanian wolf

One of the most striking similarities between the marsupials and their placental equivalents concerns the Tasmanian wolf (*Thylacinus cynocephalus*) and the true wolf. The true wolf has longer legs and is a faster runner, but in general body form and in the shape of the teeth and jaws the two animals are remarkably alike. The similarities have come about simply because the animals fill the same ecological niche — both are hunters of relatively large and fast-moving prey and therefore both need the same kind of equipment. This development of similar features in unrelated animals in different parts of the world is known as convergent evolution.

The Tasmanian wolf was once found all over Australia, where it hunted the large herds or mobs of kangaroos and wallabies. It declined with the introduction of the dingo, however, and

farmers finally exterminated it on the mainland because of its attacks on their sheep. The animal now maintains a very precarious hold on existence in the wilder parts of the Tasmanian bush, where it feeds mainly on wallabies and birds, together with various other small mammals. It does not have the social behaviour of the true wolf, but it often hunts in pairs. Its nose is probably its main sense organ, enabling it to pick up a trail and follow it in a dog-like manner. Although not particularly fast, the Tasmanian wolf has remarkable stamina and it wears down its mammalian prey in a relentless chase.

The final assault comes when the prey tires, with the wolf leaping onto it and killing it very quickly with the large canine teeth. The animal is said to lap up the blood of its victims as well as to slice up the flesh with its cheek teeth.

Small prey is disposed of completely, but larger victims are only partly eaten and, unlike many placental carnivores, the Tasmanian wolf does not return to a kill to finish it at a later date.

The Tasmanian devil

The Tasmanian devil (*Sarcophilus harrisii*) is the nearest marsupial equivalent to the bear, although, with a weight averaging only about 8kg (17·6lb), it is much smaller than any of the bears. Like the Tasmanian wolf, it was once found throughout Australia, but it could not compete with the dingo and it is now almost certainly confined to Tasmania, although it is much more common than the Tasmanian wolf and not in any danger of extinction.

The devil lives in forests and in rocky places and, like the Tasmanian wolf, it is largely nocturnal. It feeds mainly on

Nearly extinct killer
Left The Tasmanian
wolf is extremely rare,
but scattered reports
suggest that it still
exists in the wilder
parts of Tasmania.
Little is known of its
hunting habits, but its
similarity with the
true wolf indicates that
it must have very
similar habits. It uses
its nose to track down
birds and various
kinds of mammals.

Small but tenacious
Right The narrow-
nosed marsupial mouse
(*Planigale tenuirostris*)
is the smallest of the
marsupials, but it is a
pugnacious creature.
Here it is feeding on a
grasshopper as large
as itself, chewing it
up with numerous
sharply-pointed teeth.
The animal lives in the
dry grasslands of
Australia.

birds, lizards, and the smaller wallabies and rat kangaroos, which it pounces on and kills with its formidable teeth and claws. It sometimes pulls down quite large wallabies, and it also eats fishes and frogs, wading into the water to catch them or flipping them out with a paw just like brown bears (see page 106). Carrion is readily eaten, and the devil probably fed regularly on the Tasmanian wolf's leftovers when the wolf was a more common animal. Smell and hearing are the two main senses involved in finding food.

Marsupial cats and mice

The marsupial cats, typified by the tiger cat (*Dasyurops maculatus*), are rather more like martens or genets than true cats. Ranging from the size of a weasel to that of a domestic cat, they have short legs, a long tail, and a pointed muzzle full of very sharp teeth. All five species climb well and spend a lot of time in the trees. The tiger cat has rough pads on its feet to give it extra grip. Birds and their eggs, insects, lizards, rats, mice, and fish all figure prominently in the diets of these marsupial cats, or dasyures as they are often called. The tiger cat, the largest of the species, also catches some of the smaller wallabies, and all of the dasyures have been known to cause extensive damage on poultry farms. Movement stimulates them to chase and kill and, like the weasel and fox, they often kill far more than they can eat when surrounded by captive chickens.

The dasyures are primarily nocturnal and they use their large eyes to detect prey. They stalk it in cat-like fashion until near enough to make the final leap. When hunting in the trees, the animals often make tremendous leaps from one branch to another to snatch a bird or other prey. Their amazing agility enables them to cling to the branch and the prey as a rule, but even if they crash to the ground they generally manage to land on their feet, with the prey held firmly in the jaws. Like cats and weasels, the dasyures normally kill their prey with a deep bite on the back of the neck.

Australia's marsupial mice have almost nothing in common with true mice, apart from small size and scampering habits. They are actually related to the dasyures and are largely insectivorous. It would perhaps be more sensible to call them marsupial shrews, especially as they all have shrew-like snouts. There are about 30 species, ranging from the size of a small mouse to that of a rat. They occupy a wide variety of habitats, from desert to dense, humid forest, and are active mainly by night.

Some of the larger species eat birds and other mammals, and they have amazing appetites. True mice are a favourite food of some of these marsupials. They are detected and followed largely by ear, although scent and sight probably play a part, and the marsupials' long whiskers must help them to find some of their insect food. Most prey is pounced on and killed with a bite on the neck. Mammalian prey is often cleverly skinned before it is eaten; the predator begins at the snout and systematically rolls back the skin as it munches the flesh and bones. All that is left at the end of the meal is the inside-out skin — and a rather plump marsupial. Some of these little creatures have been known to eat more than their own weight of food in a night. The fat-tailed marsupial mouse stores food in its tail and draws on this store in hard times.

The Tasmanian devil
Left The Tasmanian
devil is the marsupial
equivalent of the bear,
although it is much
smaller than any bear.
It is a powerful animal
for its size, and it can
bring down wallabies
much larger than
itself with the aid of
its strong claws and
formidable battery
of teeth.

A firm grip
Right The tiger cat
spends a lot of time in
the trees. It can make
tremendous leaps from
one branch to another,
and rough pads on its
feet assure it of a
firm grip. Its large eyes
give it excellent vision
for hunting at night
— both in the trees
and on the ground.

The Anteaters

Many birds eat ants and termites when they get the opportunity, especially when these insects emerge in their millions for their mating flights. The insects usually do no more than contribute to the birds' overall diet, however, and it is only among the mammals that we find predators specializing in ants and termites. The individual insects are, of course, very small, but they live in colonies with perhaps millions of individuals packed into a small space. Such larders are very well worth breaking into.

The nature of the prey

Before describing the predators, it is worth a short digression to look at the ants and termites themselves, for these insects are often confused with each other. Termites, for example, are often called white ants. They look and behave like ants in many ways, but the two groups are really very different.

The termites belong to the order Isoptera and are closely related to the cockroaches, which are among the most primitive insects. The colony is ruled by a king and a queen, and the work force consists of immature termites of both sexes. Most of these juveniles never grow up completely, but a certain number become mature each year and go off to mate and possibly start new colonies. Almost all termites live in tropical and subtropical regions, their nests ranging from simple tunnels in dead wood to huge mounds of soil with the hardness of concrete.

Ants belong to the order Hymenoptera and are related to the bees and wasps. Each colony is ruled just by a queen, and the work force consists of adult, but sterile females. Ants live almost everywhere and make a wide variety of nests, but their mounds are never as large as those of termites.

Most of the so-called anteaters should really be called termite-eaters because they eat far more termites than ants. Termites have softer bodies than ants and they are not armed with stings, although the soldiers have powerful jaws. Termite-eating has evolved in several quite distinct groups of mammals, but all have similar equipment for opening the nests and dealing with their tiny, but very nutritious, victims.

The giant anteater

The giant anteater (*Myrmecophaga tridactyla*) and its two relatives, the silky anteater and the tamandua, live in South and Central America. They are all members of the mammalian order known as the edentates, a name which means 'toothless', although the other members of the group – the sloths and armadillos – do have some teeth.

The giant anteater lives in open forest and on the grasslands, where the ground is dotted with large and small termite mounds. It is active by night or day. It has small eyes, but it finds its food almost entirely through its keen sense of smell, which is located in the extraordinarily long snout. The tip of the snout is always applied to the ground as the animal moves about, and it easily picks up the scent of ants or termites. The anteater is thus led to the nest and it uses the huge claws on its front legs to rip open even the hardest mounds. These claws are so large that the anteater ambles along on its knuckles with the claws turned inwards so that they do not get in the way or become damaged.

The other use of the long snout then becomes apparent: it is thrust deep into the nest and the tongue, up to 25cm (10in) in length, is used to lap up the insects. It is covered with very sticky saliva and it traps the insects like a flypaper traps flies. Up to 500 may be captured with one lick, and the anteater can eat many thousands in a day. Thick skin protects the animal from the bites of the termite soldiers, but we do not know how it copes with the stings of ants; it may be immune to the effects of the venom, or its skin may simply be too

tough for the stings to penetrate. The soft tongue seems very vulnerable to irate insects, though the thick layer of viscous saliva may provide some protection. The soft-bodied termites are easily digested, but powerful stomach muscles are needed to grind up the hard-bodied ants.

The other two anteaters are arboreal creatures with prehensile tails. They raid the smaller, but very numerous ant and termite nests on the trunks and branches of forest trees.

Scaly anteaters

The seven species of scaly anteaters, or pangolins, (*Manis* spp) live in the tropics of Africa and Asia. They belong to the order Pholidota. Except on the underside, their hair has been converted into sharp-edged, overlapping scales, and the animals can protect themselves against almost any adversary by rolling into a ball. Like the other anteaters, the pangolins have long snouts, long, sticky tongues, and strong claws for ripping open ant and termite nests. They have no teeth and often swallow small pebbles that help the muscular stomach to grind up the ants. Some pangolins live on the ground, while other species inhabit the trees. The arboreal ones feed mainly on

The giant anteater
Above left The giant anteater ambles along with its long snout permanently close to the ground to pick up the scent of its insect prey. This may lead it to a mound, which it can rip open with its massive claws, as shown in the diagram on the left. The tongue, 25cm (10in) long, can probe into the mound.

Clawed shovels
Above The echidna, or spiny anteater, has great shovel-like feet with which it digs out the termites. It obtains nearly all of its insects from subterranean nests.

Africa's earth pig
Right A young aardvark displays its long snout, which houses a most efficient sense of smell.

ants, and it seems likely that all pangolins eat rather more ants than other anteaters. They are certainly well equipped to deal with them, with very tough skin on the snout and eyelids, and ears and nostrils that can be closed to prevent entry if the ants get too agitated.

Australia's anteaters

Australia has a very high population of termite species and two of its native mammals have adopted termite-eating habits, although they are less specialized than the species already described. The numbat, or banded anteater, (*Myrmecobius fasciatus*) is a marsupial inhabiting the eucalyptus forests. It uses its moderately long snout to sniff out termite nests in rotten wood or under the ground, and then rips them open with its claws. The sticky tongue is about 10cm (4in) long and it is flicked out in all directions to pick up the insects. These are then lightly crushed by the small teeth. Some ants are eaten, but these always seem to be species that share the termites' nests.

The spiny anteaters, or echidnas, (species of *Tachyglossus* and *Zaglossus*) are monotremes, or egg-laying mammals. They eat a wide variety of insects, but ants and termites make up a large proportion of the diet. They are detected by smell as an echidna scuffles along the woodland floor, using its shovel-like front paws, equipped with massive claws, to dig them out. Like all the other ant-eating mammals, echidnas pick up their prey with the aid of a sticky tongue. There are no teeth, but the insects are shredded between horny ridges on the roof of the mouth and spines at the base of the tongue.

The aardvark

The aardvark (*Orycteropus afer*) is the world's most highly specialized termite-eater — so different from any other animal that it is given an order to itself — the Tubulidentata. It occurs in most parts of Africa south of the Sahara, except for the equatorial forests. Its name is an Afrikaans word meaning 'earth pig.'

The aardvark is a nocturnal animal and, although it has reasonably large eyes, it relies on its sense of smell to find food. The long snout has a very extensive olfactory surface inside and, together with a very large olfactory region in the brain, this gives the aardvark a most efficient sense of smell. Guided by its nose, the aardvark digs termites out from rotten wood and from their underground tunnels, but its main targets are the large termite mounds. It sits back on its haunches in a kangaroo-like posture and attacks the mounds with the stout claws of its front feet. Its legs and feet are extremely strong and it can dig into mounds that defeat men with pick-axes. The excavated material is scraped away with the hind feet.

As soon as the outer wall has been breached, the termites swarm out to investigate and they are licked up by the aardvark's long, sticky tongue, which can also probe some 45cm (18in) into the mound. The animal has no teeth at the front of its snout, but there are twenty rather curious, peg-like cheek teeth that lightly crush the termites. The aardvark seems unable to deal with ants and lacks the muscular stomach found in the giant anteater and the pangolins.

Like the other termite-eating mammals, the aardvark has a tough skin to protect it from the bites of the soldiers. It also has a stiff fringe of bristles to keep the insects out of its nose, and it can close its nostrils completely if the insects become too troublesome.

As well as providing the aardvark with food, the termites may also provide it with a home; after digging out part of a mound and eating most of the insects, the predator often curls up and goes to sleep in the centre of the termite nest.

Hunters in Concert

Two heads are better than one – if only because they contain more teeth. Several predatory species work on this principle and hunt in packs: community spirit ensures that shared efforts bring shared rewards.

Most of the hunters described in this book are solitary creatures, hunting alone and purely for themselves or for their immediate offspring. As soon as the youngsters grow up they have to make their own way through the web of life, and many are on their own from the moment they are born. A few predatory animals, however, indulge in some kind of group hunting. The association is sometimes a very loose one, as in the pelicans (page 184) and the thresher sharks (page 186), but there are some hunters that always live in well-regulated and tightly-knit social groups. They are all mammals or social insects, and the social unit is always an extended family group. We have already seen how cheetahs sometimes stay together when they are grown up (page 110), and it is not difficult to see how a full social life could have developed from such family orientated beginnings.

Numerical strength

In the same way that many herbivorous animals gain safety in numbers, so the social hunters gain strength from their numbers. By working together, they can capture much larger prey than they could if working singly, and the return per unit effort is generally much greater. New techniques are also possible with group hunting because a team of hunters can split up and attack prey from two or more directions. Lions, for example, often encircle their quarry and those on one side move in to drive it into the jaws of the lions waiting on the other side. Hunting dogs (page 134) use techniques that verge on relay-running to capture antelopes, while several other predators work in pairs or small groups to separate young ungulates (hoofed mammals) from their mother so that the young can be captured without trouble.

Group hunters also benefit from the different ability levels developed by members of the community. One individual may be particularly fast over the ground, while another may be better at detecting prey than chasing it. Combining these talents into a single operative unit is obviously of great value to the community as a whole. It is the beginning of the division of labour. Wolves and hunting dogs share the jobs very efficiently; one group goes off to hunt, while another stays behind to look after the cubs, although it is not always the same individuals that do each job and the division of labour is not clear-cut. Among

the lions, it is the lionesses that do almost all of the hunting, while the males defend the territories, but it is among the social insects (see below) that the division of labour is most marked.

Share and share alike

Group hunting and work-sharing is clearly very beneficial, but it can work only if the animals concerned can share the spoils as well; the hunters must take food back to the babysitters and babies even if it means forgoing a full meal themselves. Such behaviour is abnormal in most animals and it is sustained in the social species only by means of elaborate systems of communication between the members of the group. The greeting and pre-hunting ceremonies performed by wolves and hunting dogs are clearly components of these communication systems, designed to maintain the community spirit in each pack. The mammals learn much of this behaviour while they are young and in the care of their parents, but the insects do it all by instinct.

The social insects

Social insects belong to just two of the 30 orders of insects – the Isoptera, or termites, and the Hymenoptera, which includes the bees, wasps, and ants. Termite colonies are 'ruled' by both king and queen, but the other social insects are led purely by queens – usually just one queen to each colony, although she may have several million subjects. Most of these are workers – sterile females who do all the chores from building the nest to feeding the babies, which are, of course, their younger sisters. Males appear in the colonies only at certain times, their sole function being to mate with poten-

tial new queens that are also reared at these times. Many ant species have several different kinds (castes) of worker, including some extra large ones known as soldiers that defend the colonies. Each kind of worker does a specific job, and the division of labour is thus very well defined within the colony.

Control of the colony is exercised by way of pheromones, or scent signals, produced by the queen. These permeate the whole colony and trigger off instincts in the various castes so that they perform their correct duties. Removal of the queen leads to an immediate breakdown of law and order, but things work so smoothly and efficiently when she is in residence that William Wheeler, a noted American entomologist, was led to compare the whole colony to a single body and to liken each individual member to the cells of that body. This is known as the super-organism theory and, although biologists no longer pay much attention to it, it cannot be denied that a colony of social insects, particularly army ants, does behave very much like a single organism.

Among the predatory social insects, only the army ants hunt in concert. The social wasps hunt and retrieve food individually, and most ants also hunt individually, although they follow each other along trails leading to the hunting grounds. Each ant brings back its own booty, although several ants may be seen tugging at a large insect such as a butterfly, each unaware that sister ants are tugging at it from different positions. Army ants, however, always search in groups and they combine to carry their booty back to the nest – co-operative hunting at its best and most effective.

The concerted attack
Left Army ants regularly invade the nests of other species of ants and destroy both adults and young. Here a group of army ant workers is attacking a soldier from a leaf-cutter ant colony in Trinidad.

A strong pair bond
Right Coyotes do not hunt in packs like the timber wolves, but they have strong pair bonds and male and female often hunt or scavenge together, as they are doing here. They sometimes combine to drive prey into an ambush.

Armies of Ants

There are more than 10,000 known species of ants, all of them social insects living in colonies ruled by one or more queens. A mature colony may contain anything from a few dozen individuals to more than 20 million, depending on the species. Some species, such as the fungus-eating leaf-cutter ants and the seed-eating harvesters, are purely vegetarian, but the majority of ants are omnivorous and eat other insects just as readily as they lap up fruit juices or honeydew from aphids.

At the other end of the scale, there are some purely carnivorous ants. The most famous of these are the 240 or so species of army ants, which are defined in at least one dictionary as '. . . ants that go out in search of food in companies. . . .' Edward Wilson of Harvard has actually described an army ant colony as '. . . an animal weighing in excess of 20kg (44lb) and possessing in the order of 20 million mouths and stings . . . surely the most formidable creation of the insect world . . .' His reference to the colony as an animal is in line with William Wheeler's 'super-organism' theory (see page 120) and indicates the amazing co-operation that exists between the individual ants; they really do seem to act as one flowing organism with a single control centre.

The awesome army ants

Army ants, also known as legionary ants and driver ants, eat virtually any animal matter that they can lay their powerful jaws on. They make an excellent job of clearing rats and other pests from houses; any animal that cannot escape is eaten alive and, although meat may be eaten in the pantry, no other damage is done to the property and the owners can return as soon as the ants have left. A tethered leopard has been reduced to a skeleton within hours, and pythons weighed down by heavy meals have suffered a similar fate, but the army ants' main victims are smaller creatures that cannot get out of their way. Other insects form the bulk of their food under normal conditions, with the nests of bees and wasps being especially vulnerable. The ants stream through these nests, putting the adults to flight and destroying every single grub. The nests of termites and other ants may also be attacked, and the warriors may even take over the nests to provide them with temporary shelters.

The army ants do not have settled homes like the other species. With populations reaching 20 million or more in some colonies, they would very soon exhaust food supplies if they stayed in one area for long. They rarely remain in one place for more than about three weeks, the move often being timed to coincide with the hatching of a new

batch of eggs. All army ants are tropical in their distribution, and it is doubtful if they could survive anywhere else because food supplies would be insufficient.

American army ants

One of the best known species of army ants is *Eciton burchelli* from tropical America. Its colonies contain up to 700,000 individuals. During the residential phase, which is the time of egg-laying, the whole colony occupies a bivouac site in a sheltered spot. The site is often under a fallen tree or between the buttress roots of a standing trunk, but the ants make no other shelter. The workers lock themselves together with their feet and, with the queen and the eggs and pupae in the centre, they form a huge ball up to 1m (39in) in diameter. The 'nest' is thus composed of the workers' bodies.

At daybreak, when the light level reaches 0·05 foot candle (see page 144), the bivouac begins to disintegrate and the ants surge in all directions. Eventually they 'decide' on one direction and the day's raiding party sets out like a

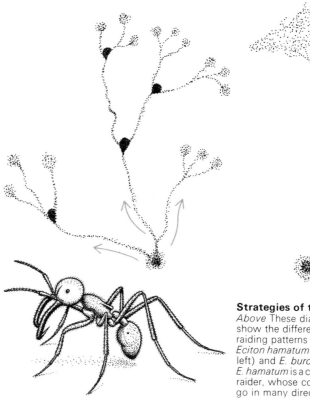
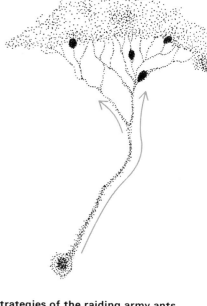

Strategies of the raiding army ants
Above These diagrams show the different raiding patterns of *Eciton hamatum* (above left) and *E. burchelli*. *E. hamatum* is a column-raider, whose columns go in many directions. *E. burchelli* is a swarm-raider, forming a broad front. Both species cache food at various points along the way (dark patches). On the left is a huge-jawed *E. hamatum* soldier.

Forest butchery
Left A group of *Eciton* workers have over-powered a bush cricket and are preparing to cut it to pieces with their jaws before carrying it back to their bivouac.

A living safety rail
Right A column of African driver ants (*Anomma nigricans*) streams across a rough forest path. The large-bodied soldiers form a living wall to prevent the workers from falling into a gulley cut by rain water running down the path.

A sizeable meal
Below A large beetle grub has been found by a worker driver ant of the genus *Dorylus*. Undaunted by its giant prey, the ant begins to carve it up. Other ants will soon join in.

thick brown rope snaking its way over the ground at a speed of perhaps 20m (65·5ft) per hour.

There are no commanders in this army of blind warriors and the lead is continually changing, but the insects are well organized and when once the direction has been settled they deviate very little from the path. The soldier castes march at the sides of the column and, as in other ants, they defend the workers with their great jaws. They also help the column to overcome obstacles, such as deep cracks in the ground, by forming bridges and safety barriers with their own bodies.

When the ants are perhaps 20 or 30m (98ft) from the bivouac they start to fan out and form a broad front, up to 20m or so across. More ants pour in from behind and the raiding swarm presses forward. The plunder begins in earnest as all manner of insects and other invertebrates are stung or bitten to death. The noise produced by millions of tiny feet trudging over the ground is surprisingly loud, and it is augmented by the buzzing of the numerous flies and other insects

that take to the air to escape from the marauders. The chorus is also swollen by the calls of ant-birds that flutter around the army and snap up many of the insects that are put to flight. The approach of the army is certainly well advertised by this commotion.

Much food is eaten on the move, but a good deal is also cached along the route. Forward movement ceases after a few hours and, although ants may still be streaming from the bivouac, the main raiding party begins to retrace its steps, following the strong odour trails laid down on the outward journey. The food stores are collected, cut up if necessary, and carried back to the bivouac to be shared with the queen and the other ants that stayed behind to guard her. Another raid takes place the next day, and this goes on until the eggs start to hatch and the pupae stored in the bivouac produce new workers.

The population increase marks the start of the nomadic phase. Activity in the bivouac increases dramatically, and when the raiding party returns the whole population moves out, carrying the larvae with them. This movement begins round about sunset and the ants take several hours to settle into a new bivouac perhaps 100m (328ft) away. This is the pattern for the next two or three weeks; a raiding party goes out each day and then the whole colony moves to a new bivouac in the evening. The ants become 'residential' again as soon as the larvae start to pupate and the demand for food drops. T. C. Schneirla, who devoted much of his life to studying ants, was the first to discover this link between the life cycle and the nomadic and residential phases of activity.

Other *Eciton* species in America have similar behavioural patterns, but *E. burchelli* is unusual in swarming on a broad front. Most of the army ants forage in relatively narrow columns that move in several directions from the bivouac.

African army ants

The various species of *Anomma* from tropical Africa are generally referred to as driver ants. They form broad raiding swarms like those of *Eciton burchelli*, but their emigrations are much less regular. The residential phase may last from a few days to three months, and the bivouac is a much more elaborate affair than that of *Eciton* and other army ants. The workers dig out a vast network of underground tunnels, which may extend to a depth of 4m (13ft) in a long-established bivouac, and the raiding parties go out from there. They usually go out at night, but they may leave at any time.

The queen seems to lay eggs more or less all the time, although she has bursts of extra productivity every two or three weeks, and the emigrations do not seem to have any links with the larval development. When the ants do emigrate, they move in a long line and they keep going day and night. The new bivouac site may be only a couple of hundred metres away, but it can take several days for the 20 million or so ants to complete the move.

Anomma has much more strongly toothed jaws than *Eciton* and the workers can more easily strip down a large vertebrate to its skeleton. *Eciton* rarely does this, although it will certainly kill any vertebrate unable to get out of the way of the raiding swarm.

The Sociable Lions

Frequently known as the king of beasts, the lion (*Panthera leo*) is the undisputed top predator of the African savanna. Along with the tiger (see page 108), it is the largest of the cats and, if we exclude the omnivorous bears, it is also the largest of all the land-based carnivores. Adult males may reach 2m (6·5ft) in length, excluding the tail, and may weigh as much as 230kg (506lb). Females, or lionesses, are on average about two thirds of this size.

The pride

Unlike other cats, the lion is a sociable creature, with a very well developed social life. The social unit is known as the pride, and it may contain between three and forty individuals, although twelve to fifteen is a more usual figure.

Most of the members are lionesses and cubs of various ages. There are rarely more than three adult males in each pride. The females are usually all related to each other, having grown up together in the pride. The males may also be related to each other, although they are not normally related to the females. The latter rarely leave the pride into which they are born, thus forming a continuing unit of sisters and cousins.

Young males leave their prides when they are between two and five years old and they go off in small groups to lead nomadic lives for a while. When they are fully mature they 'take over' other prides by driving out the older resident males. This replacement of male lions occurs every three or four years and ensures a mixing of genetic material.

Within each pride there is a good deal of 'friendly' activity, with the lions rubbing against each other with clear signs of enjoyment. Such contact may well be a way of exchanging and spreading scent secretions that help to bind the pride members together in a spirit of co-operation. The success of the pride depends upon such co-operation and the friendly relationships between the lionesses, which all seem to be equal in rank. Squabbling does occur when the animals are jostling for position at a kill, but this is rarely serious.

All the mature females can mate and breed, in contrast to the situation we find among the wolves (see page 130), and all share in feeding and protecting the cubs. A lactating female will suckle any of the pride's cubs, although she prefers her own. The mixing of the cubs at an early age is an essential factor in the stability of the pride.

The adult males, although an integral part of the pride, rarely mix with the females unless the latter are in mating condition. Apart from mating, the prime function of the males is the defence of

the pride and its territory. This can take them some distance from the body of the pride, and even when the animals are resting in the daytime the males often lie up in a group several metres away from the females and cubs.

The lion's habitat

Lions can be found throughout the savanna regions of Africa, but they are not common in either the treeless regions or the densely wooded parts. They prefer those areas with scattered trees and patches of scrub. The latter provides cover for hunting, and the trees provide essential shade for daytime slumber. In some areas the lions actually rest in the trees, but such behaviour is

frequent only where the trees are large and provide comfortable branches.

A varied diet

A list of the various animals killed and eaten by lions would be very long indeed, for lions, like most other large carnivores, are opportunist hunters and they chase, although they do not always catch, almost anything that presents itself. The menu thus depends largely on what is available.

Nevertheless, lions do have their preferences and over most of their range the bulk of their kills consist of wildebeeste and zebra. These ungulates are picked out even when others, such as Thomson's gazelle, are much more

The search
Left The agile lionesses are the most successful hunters. One may spot prey and initiate the hunt, but the more lionesses that join in the greater are the chances of success.

The chase
Top Perhaps 40 percent of hunts are successful when lionesses join forces, but a single hunter stands only a 15 percent chance of success. The species of prey caught varies with the season. Small prey like this warthog are popular quarry during the dry season.

The kill
Right These lionesses have killed a harte-beeste and dragged it to the shade to enjoy a leisurely meal.

sliced up by the
teeth. Nearly every
though the skin is
with most cats, the
is discarded. The bor
by the extremely ro

Allowing for the
prey, about 5kg of
something over two
be killed each year
lioness. This represe
or 16 wildebeeste
practice the menu w
smaller animals.

If food is scarce,
during the dry seas
the prey animals mi
tures, the adult lion
without food for a

they can endure th
hardship and make u
is made. They can take
of their body weight a
they are hungry, endir
distended belly and be
lethargic than usual.

The cubs, howeve
starvation and many
food is in short sup
question of the adults
the sake of the cubs,
detrimental to the pri
the adults became to
there would be no pride
readily replaced and so
able in the interests of

abundant. For example, about one
quarter of the lion kills witnessed or
discovered in Nairobi National Park over
many years have been wildebeeste,
although this species makes up only
about seven percent of the local large
mammal population. But we do not
know quite why this species is selected.
It may be that the lions really do prefer
the taste of their meat, but it may
equally be that the wildebeeste are
easier to catch than the other animals,
or that the lions choose them because
they realize that the wildebeeste provide
more food than the smaller antelopes.

On the other hand, none of these
suggestions may be true, and it may be
the inbuilt bias of human observations

that accounts for the apparent domi-
nance of wildebeeste and zebra in the
lion's diet. These larger kills take
longer to dispose of than smaller kills
and have a much greater chance of
detection by human observers.

Other animals that contribute appre-
ciably to the lion's diet include topi,
eland, hartebeeste, impala, various gazelles, and warthog. Those taken in
smaller numbers include the young of
much larger creatures, such as elephant,
giraffe, rhinoceros, and hippopotamus.
Buffalo are also eaten, and here a fair
number of adult males are taken in
addition to the calves. This seems
surprising at first, but mature males
tend to be solitary and thus not too

difficult for a group of lions to tackle.

Many smaller animals, such as hares,
rodents, monkeys, and birds, are eaten
when lions stumble upon them, but the
lions rarely bother to chase such small
prey. True hunting behaviour is directed
only at the larger prey species.

Like the other large cats, lions are
not averse to eating carrion and they
scavenge what and when they can,
often watching for descending vultures
that might lead them to a meal. They
will rob hyenas and other carnivores of
their kills as well. Brian Bertram, who
studied lions in the Serengeti National
Park for four years, calculated that
carrion formed about 14 percent of the
lions' food in that area.

The Voracious Wolf Pack

The wolf (*Canis lupus*) is the largest member of the dog family (Canidae) and, although man has now exterminated it in many places, it still ranges over a very large area of the northern hemisphere. Its range is probably greater than that of any other land mammal apart from the rats and mice transported by human agencies. It is equally at home in lowland forests, on rugged mountains, and on the windswept tundra of the Arctic.

Wolves live in packs, which are generally family groups with no more than about eight members, although packs of about 15 wolves are not uncommon and packs with more than 20 members have occasionally been recorded. Each pack is led by a dominant male, which is the most aggressive individual in the group and the one that decides when and where to hunt. Most of our detailed information on wolf behaviour comes from North American studies, particularly those of Adolph Murie in Alaska and David Mech on Isle Royale in Lake Superior.

Summer and winter diets

During the summer, wolves eat quite a wide range of animals, including Arctic and snowshoe hares, marmots and other rodents, lizards, fishes, and even grasshoppers and other insects. They also eat small amounts of fruit, but none of these foods is taken with any regularity; they are not sought out, merely picked up when they come to the wolf's notice. The smallest prey taken on anything like a regular basis is the beaver.

These assorted animals account for less than half of the wolves' summer diet in North America; the bulk of their food consists of deer and other hoofed mammals. Many of these are larger than the wolf itself, and it is the pack system that enables the wolf to prey on them. Observations on Russian wolves suggest that they eat a rather higher proportion of small mammals, but this is merely because large species are less common than they are in America. The wolves always go for the largest prey they can get, for large prey gives a much better return in terms of weight of food obtained for a given expenditure of energy.

Most of the small animals stay underground during the winter, and then the wolves obtain almost all of their food from the hoofed mammals. Reindeer, or caribou, moose, wapiti, and various smaller deer such as the white-tailed deer and the mule deer are the main prey animals in North America, but only one or two of these are normally present in any number in a given area. The wolves concentrate on the most easily caught species in each region. A few bison are killed in some places, while bighorn and

Dall sheep and Rocky Mountain goats are eaten in mountainous regions. Some musk oxen are taken in northern areas. Domestic sheep and cattle are readily accepted, and the wolves are not averse to eating carrion, including dead members of their own kind.

The following paragraphs deal mainly with the wolf's methods of hunting hoofed prey. Organized hunting is always a daytime occupation, but, as we shall see, many of their hunts are unsuccessful and the wolves are therefore almost permanently on the lookout for prey to satisfy their hunger.

A good nose

Wolves sometimes go hunting alone, especially in the summer when there are young deer to be had, but most of their hunting is done in packs. The pack may come upon prey as a result of a chance encounter in the forest, but most of their food is found with the aid of the nose, which, as in all dogs, is incredibly sensitive. The wolf's hearing is also very keen. Its ears are sensitive to a much wider range of frequencies than our own, and it is believed that a wolf can recognize the howls of its pack-mates from at least 6km

(3·75 miles) away. Eyesight is good, especially for detecting movement, but the eyes do not play an important role until the later stages of the hunt.

Wolves occasionally track down their prey by picking up a scent trail on the ground and following it, but the usual method of detecting prey is by picking up its scent in the air. Most of Mech's observations of wolves and moose indicate that the wolves pick up the scent when they get to within about 300m (984ft) of the prey, but a pack was once seen to pick up the scent of a moose and her two calves from a distance of about 2·5km (1·6 miles).

Picking up the scent leads to a very distinct sequence of behaviour, which begins with all the pack members coming to a halt and raising their heads to sniff the air very deliberately. They turn to face the source of the scent, and then there may be a kind of pre-hunting ceremony, with a good deal of tail-wagging, before the wolves embark on the next stage – the stalk.

The aim of the stalk is to get as close to the prey as possible and, although there is still much tail-wagging and obvious excitement, the wolves keep

---

Hunting st
Almost all
by the lion
hunts take
can run at
(34mph),
distances.
easily outr
must empl
niques to o
aids them ir
helps to co
there is littl
are the mai
of the eye h
tapetum, bel
in dim light
smell is po
little part in

One lione
better huntr
of her pride,
of hunting o
a potential
towards it,
prey is not
stantly if it l
to the prey, s
the stalking p
Using all av
forward with
she is lucky,
distance of h

Communal
all the pride
gether, is mo

Bold marauders
Above Wolves will eat carrion when they can get it, and even try to drive the powerful grizzly bear away from a mule deer carcase.

Briefly satisfied
Above right A well-fed wolf stands over a deer carcase after a good meal. Digestion is quick, and the wolf may feed again in an hour.

very quiet and peer intently ahead as they move forward in single file. They approach from the downwind side automatically, having picked up the scent and turned towards it, but they rarely get within about 10m (33ft) of the prey before they are detected. The next phase of the operation depends upon the reaction of the prey.

Confront or flee
An adult moose is quite able to defend itself against wolves, and it will not infrequently stand its ground and confront its attackers when it sees them. By kicking out with its powerful hooves, it soon persuades the wolves to move on in search of easier game. But not all moose confront the wolves to start with; only the fittest and most confident ones do this. The others turn and run as soon as they sense the approaching wolves, and this in turn releases the chase reaction of the wolves.

A fit moose can easily outrun wolves on good ground and many get clean away, for the wolves do not continue the pursuit unless they make significant gains during the first few seconds. Even if the wolves do catch up with a running moose the chase is not necessarily over; the moose may turn and confront the wolves, or it may go on running and eventually out-distance them.

Mech witnessed many stalks on Isle Royale – where the moose is virtually the wolf's only prey – and found that 96 out of 120 adult moose turned and ran as soon as they sensed the stalking wolves. The other 24 stood their ground and came to no harm. Of those that fled, 43 got clean away. The wolves caught up with the other 53, of which 12 then turned and confronted the attackers to good effect and 41 continued to run. Only seven were unable to outrun the wolves in the end, and a mere six moose were finally killed.

This low success rate shows very clearly why the wolves have to spend so much time hunting, although they certainly have more success with moose calves. The latter are generally with their mothers, but a pack of wolves is usually

able to worry the mother sufficiently for one or more of their number to be able to sneak in to take the calf.

Adult wapiti, or elk, may react in the same way as the moose, but the smaller deer generally flee straight away, whether they are fit and strong or not. The wolves give chase, but again they do not waste their energy if it becomes obvious that they are not going to catch the deer within a short distance. It really seems as if they are setting the deer a sort of survival test and punishing those that fail with death. The failures are usually the very young, the old, and the sick animals in a group.

Wolf versus caribou
The caribou tends to live in large herds, whereas moose are more often found alone. The caribou also has greater stamina and can run for long periods. Caribou-hunting therefore requires a different approach from that used against the moose, and the wolves use three different strategies. Finding the caribou is not difficult, and many caribou herds actually have 'resident' wolves following them nearly all the time. These do not seem to worry the caribou unless they

get too close, and here is the wolves' problem – how to get close enough to make a successful capture.

The ambush is one method that is used, especially when a small group of caribou is involved – perhaps a group that has wandered away from the main herd. One or more wolves move to the far side of the caribou, taking up position on the uphill side of them if the terrain allows, and the rest of the pack then drive the caribou into the trap. Single wolves also ambush caribou on occasion, hiding themselves uphill of the quarry so that they can rush down on the prey when it comes within range.

Some observations suggest that the wolves use a relay system to capture caribou, but this may be just a chance happening. One can imagine part of a pack of wolves chasing after some caribou and leaving the rest of the pack behind. If the caribou then curve round and come close to the resting wolves it is easy to believe that they might take up the chase and give the original pursuers a rest, but this may not imply that the wolves are working to a plan. Only if the wolves continued to chase the caribou round in circles could this be considered a true example of hunting by relay, and there is little evidence that this actually happens.

The most common method of attacking caribou herds is the method used by most other hunters; the herds are put to flight and the wolves look for weaklings or stragglers on which to concentrate their efforts. The weaklings might not show up immediately, and the wolves keep up the chase for a couple of minutes or so, but they rarely go on any longer unless they are clearly catching up on an individual. Prolonged observations of hunting wolves suggest that the stories of wolves chasing their prey endlessly for hour after hour are probably greatly exaggerated.

The kill
Wolves do not always kill cleanly at the first attempt, especially when attacking large prey such as a moose. They snap at its rump and flanks when they catch up with it but, although they have immensely powerful jaws and teeth, the first attack may not bring the prey down. Wolves have even been dragged along at high speed with their teeth firmly embedded in the rump of a moose. Several attacks, separated by quite long chases, may be necessary before the prey finally succumbs. It is interesting that moose are almost always attacked on the rump and flank, but caribou are generally grabbed by the shoulder. The wolves on Isle Royale frequently grab their moose prey by the nose as soon as

some of them have managed to obtain a good hold on the rump.

As soon as the prey falls, the wolf pack gather round and begin to rip the body to pieces, usually concentrating on the rich steak of the rump first. The intestines and the fatty deposits around them are eaten next, and the heart and liver go quickly too. Very large chunks of flesh are swallowed, and the wolf's rough tongue enables it to clean small scraps of flesh from the bones as well. A wolf can eat 9 or 10kg (20–22lb) of meat at one sitting, and a deer weighing 100kg (220lb) is rapidly reduced to a skeleton by an average pack. Most of the meat is eaten on the spot, but pieces may be carried back to feed females looking after cubs at the den. Some food may also be regurgitated to feed the females and youngsters.

When very large prey has been killed, and the wolves cannot eat it all at one sitting, it may be buried for another day. Caching food in this way does not hide it from the marauding wolverine (see page 96) or from foxes, but it does protect it from the ever-present ravens and, during the summer months, from the swarms of blowflies.

Caching food is not usually necessary, however, because, although the wolves can gulp down large quantities of meat at one go, their digestive processes are so rapid that they are often ready to feed again within an hour or so. A medium-sized pack can thus feed on a carcase for several hours, although no

wolf will be feeding there for the whole time. A heavy meal is usually followed by a sleep, but the wolf is then ready to feed again, and if nothing remains of the previous kill it will begin another hunt. Hunting is not always successful, however, and the wolf also has the extremely useful ability to fast for several days at a time if it fails to find food.

The coyote – wolf of the plains
The wolf described in the previous paragraphs is more accurately known as the grey or timber wolf, to distinguish it from the closely related red wolf (*Canis niger*) that lives only in a small part of the southern United States. The third member of the genus living in North America is the coyote (*Canis latrans*), which, unlike the others, has actually been extending its range in recent years. It is often known as the prairie wolf, for the open plains are its major habitat.

Like the grey wolf, the coyote mates for life, but it does not live in packs. Except when the parents are teaching their offspring to hunt, they do their hunting singly or in pairs. Jackrabbits, cottontails, and small rodents are their main food, but they will also hunt small deer and they usually rely on speed and stamina to catch them. Sometimes the male and female coyote work together to ambush the prey, but their most unusual 'trick' is to sham dead. Crows and other scavengers gather round, only to find the coyotes leaping up and grabbing them before they can escape.

Digging for food
Left Mice and voles are the coyote's main prey in most areas. They are usually caught on the surface, but the coyote sometimes digs down to reach them in their burrows, as it is doing in this picture.

Surround and kill
Right Wolves hunt in concert when chasing large prey like this caribou and calf. One wolf distracts the prey by showing itself and howling at a distance far enough not to stampede them, while other pack members circle the prey and cut off their escape. The wolves creep closer, and finally rush in for the kill. The older and more experienced caribou may escape, but the young animal is easy prey for the snarling, snapping jaws. Some wolves grasp the calf by the shoulders and others attack its nose and flanks. The prey, weakened by loss of blood and crushed muscles, soon falls.

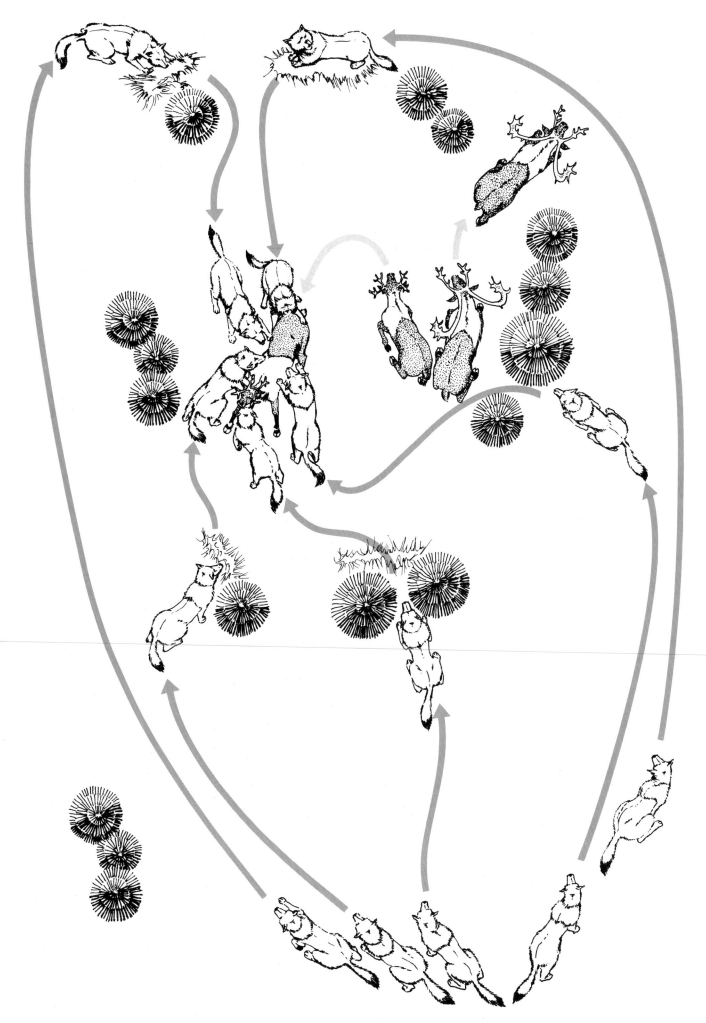

The Relentless Hunting Dog

The hunting dog pack of the African savanna is one of the most efficient killing machines in the animal kingdom, and yet it works without any apparent leader to bark orders to the ranks. Careful study of dog packs in East Africa, particularly by Wolfdietrich Kühme and the van Lawick-Goodalls, has shown that the animals actually do have a peck order—in fact, there are two hierarchies, one among the males and one among the females—but constant togetherness means that the dogs are so well known to each other that the peck order rarely has to be demonstrated. Life among the hunting dogs is one of amicable co-existence and truly amazing co-operation, both in the hunting of food and the rearing of the young.

The organization of the pack

The pack consists of anything up to about 60 animals, although a typical

No hunting in the mid-day sun
Above Hunting dogs hunt at dawn and dusk and rest during the day, well camouflaged in dry grassland by their dappled coats.

pack has between six and 20 individuals. Adult males usually outnumber adult females, often by as much as two to one, but both sexes take part in the hunt. Each pack has a distinct home range, which may extend to well over 2500km^2 (approximately 977mi^2), and the animals roam freely within this range. Only when there are young pups in the pack do the animals settle down in one place, and it has been found that the pack generally returns to the same denning area every time pups are expected.

The major foods of the hunting dog are the small antelopes, such as Thomson's gazelles, impalas, and duikers, but the exact diet varies with the area and in some places the dogs regularly kill wildebeeste, zebra, and even eland. This is some feat when you realize that the dogs are only 60cm (24in) high – half the height of a wildebeeste and only one third the height of an eland. The dogs even chase and manage to kill lions from time to time.

Well-organized team-work is the essential element of the hunt, and co-operation is ensured by way of elaborate 'ceremonies' that take place every time the animals go out for dinner. This is usually early in the morning and late in the afternoon.

The pre-hunt ceremony

Before the hunt, a ceremony is started by the 'top dog' and it involves a great deal of tail-wagging, 'kissing', and mutual licking of muzzles. This activity spreads through the group and synchronizes the moods of the individuals. Excitement builds up, and the dogs are then all ready to go off together.

The top dog may decide in which direction the pack will go, but he does not necessarily play a major role when the hunt gets under way. The animals trot along at about 10kph (6mph) until they sight a likely herd of antelope, and then, when perhaps 500m (1640ft) away, they slow down to a walk with their heads held low. In this way the dogs can get within perhaps 50m (164ft) before the herd takes flight.

On some occasions the dogs allow the herd to gallop past them without giving chase—a strange ploy which seems to enable the dogs to detect any weak antelopes in the herd and to weigh up their chances in the ensuing chase.

The chase

Usually the dogs give chase as soon as the herd starts to run. They may start by chasing several individual antelopes within the herd, but the dogs are always watching each other and as soon as they see that one of their number is gaining on an antelope they all converge on the one victim and try to cut it out from the herd. But there may still be a long chase, at speeds of up to 50kph (31mph), before the antelope is caught.

During the chase, the hunters change their positions several times, although they probably do not consciously work in relays as some people have suggested. A hunted antelope often runs in a zig-zag fashion or curves round in a wide arc, thus allowing dogs near the back of the pack to take short cuts and take over at the front of the pack. Those dogs that had been making the running at the front can then take a breather before possibly going into the lead again. In this way, the antelope is likely to tire long before the hunting pack, and the dogs eventually close in for the kill.

The leading dogs often come up on both sides of the victim, so that whichever way it turns it will run into a pair of jaws. The dogs leap at its hindquarters and bite hard and deep. The antelope very soon falls, and it is quickly torn to pieces by the excited dogs. This is not a

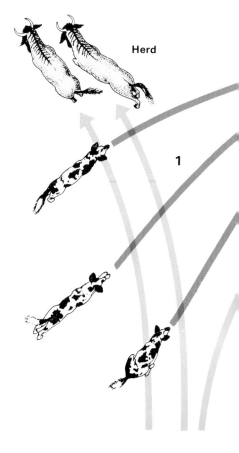

Herd

1

3

2

Potential victims
Left Wildebeeste graze in vast herds on constant alert for attack by wild dogs and other predators. At calving time, early in the year, wildebeeste make up over 50 percent of wild dogs' kills.

End of the chase
Above A wildebeeste has managed to stay on its feet after the initial onslaught, but it will very soon be torn to pieces by the dogs' great canine teeth. Despite the excitement, there is no quarrelling.

Quarry

Strategy of the chase
This sequence of drawings shows the strategy of a pack of wild hunting dogs in pursuit of a wildebeeste.

1 The dogs charge on a broad front, scattering the prey and eventually singling out a victim.

2 The victim is chased up to 3km (1·9mi). As it tires, it curves and zig-zags to evade the snapping jaws. This allows dogs on the flanks to take short cuts and gain ground. The leading dog may slow down the quarry by pulling its tail.

3 The prey is eventually brought down by a dog grabbing a leg. The other dogs quickly crowd round and attack the soft belly with great excitement. One often holds the nose until the victim is still.

pretty sight, for the animal is still alive when the dogs disembowel it, but it seems likely that the initial bites produce a state of deep shock, so that the victim does not feel further pain.

A sense of duty
Unlike the lions, the hunting dogs do not insist on the leaders eating first. In fact, they normally allow the younger ones to feed first, and the older dogs take what is left. If they have not all had their fill, the pack will kill again until they are all satisfied.

When there are pups in the denning area, some of the adults, mainly the mothers, stay behind to look after them, and the hunters then have to bring back food for the nursemaids as well as for the youngsters. They never forget, and they bolt down especially large pieces for this purpose. On returning to the denning area, the hunters are greeted excitedly by the other dogs, which nuzzle up to them and cause them to disgorge the large chunks of meat.

Old and lame dogs unable to keep up with the chase are fed in the same way, showing that the hunting dogs have a truly remarkable community spirit and sense of duty to their fellows. But perhaps the most remarkable thing is the way in which the older hunters will often pull back at the last minute and allow the younger, inexperienced animals to complete the kill. It is almost as if they know that the younger animals must learn to kill, for these are the animals that will have to support the pack in the future seasons.

thes

The only anima
quered the air a
and the birds.
different groups
evolved in very
have all mastere
and they conta
plished aeronaut

The true aer
animals that ac
air or at least us
prey. We can
major groups —
catch their food
swoop down to
the ground; and
to take their p
Examples of the
air combat speci
eating bats, drag
such as swallows
as some of the
attackers include
a few insects and
specialists includ
other seabirds, k
bats and birds of

Spotting the quar

Sight is obviously
most aerial hunte
animals have inc
A kestrel hovering
ground can easily
beetle, scuttling th
but such an obj
invisible to our
probably be unde
as well if it did not
eyes, like those of
respond primarily
why many prey a
they see a predato

It is the dense pa
cones in the retin
gives the hawks su
the more sensitive
retina, especially c
to pick out detail. In
may have up to
square millimetre in
its retina; eight tin
Hunters searchin
eye use several tec
usually hovers in th
ground below for sui
is seen, it moves t
begins again, thus c
from relatively fev
Eagles and vultu
the air on their b
round and round o

and rabbits because these animals are
very fast themselves. In very rugged
areas the eagle may hunt from a perch
on a rocky crag.

Several other eagles hunt in much the
same way as the golden eagle. The
martial eagle, for example, soars high
over the African savanna as it searches
for large game birds. It then swoops
down on them in a beautifully controlled
dive. The martial eagle is the largest of
the African eagles and it will eat a wide
variety of prey, but it definitely prefers
game birds somewhat smaller than itself.
This explains the relatively small feet of
this enormous and very handsome bird.

Eagles of the forest

Some of the bird-eating raptors of the
forest are described on page 158, but
birds are not the only inhabitants of the
forests and it is not surprising to find
that several eagles have become adapted
for catching mammals in these habitats.
The most famous, perhaps, is the harpy
eagle (*Harpia harpyja*) of tropical America,
with the rare monkey-eating eagle (*Pithe-
cophaga jefferyi*) of the Philippines not far
behind in either fame or size.

The wings of these great birds — the
largest of all the eagles — are relatively
short and very broad and, together with
the long, rudder-like tail, they give the
birds all the agility they need for weaving
through the trees. Speeds of 80kph
(50mph) have been recorded for the
harpy eagle. The birds use their eyes and
their ears to detect prey. The harpy eagle
snatches monkeys, sloths, porcupines,
opossums, and many other mammals
from the branches, and also takes agoutis
from the forest floor. Little is known
about the very rare monkey-eating
eagle's feeding habits, but it probably
eats flying lemurs, or colugos, (*Cynocephalus variegatus*) and various other
mammals as well as monkeys.

Snake-eating eagles

There are twelve species of snake eagles
or serpent eagles, all confined to the Old
World. The short-toed eagle, or European
snake eagle, (*Circaetus gallicus*) extends
into Germany during the summer, but
the other snake eagles are tropical or
subtropical birds inhabiting a wide
range of habitats. All are large-headed
birds, often with striking crests of
feathers, and all have short, strong toes
adapted for taking a firm grip on the
body of a snake. The short-toed eagle
hunts by hovering rather like a kestrel,
and the bateleur eagle glides to and fro
over its hunting grounds, but the other
snake eagles hunt from perches.

With the exception of the bateleur
eagle, the snake eagles feed almost entirely on snakes. The prey is grabbed any-

The martial eagle
Above Largest of the
African eagles, the
martial eagle (*Pole-
maetus bellicosus*)
soars high over the
savanna in its search
for prey. Game birds
are its favourite prey
but it also kills
rodents and small
antelopes like the on
on which it is feedin
in this picture.

A snake for baby
Left The male short-
toed eagle, on the
left of the picture,
has just given the
hen a snake with wh
to feed the solitary
youngster in the ne
The chick will take
the snake by the he
and swallow it who

The secretary bird
Right A secretary bird begins to swallow a snake that it has caught and trampled to death. The bird stamps on its victims so hard that they are some-times virtually skinned. Secretary birds are known to kill snakes more than 1m (39in) long, but it seems that some individuals are much more ready to catch snakes than others. Insects and small mammals make up the bulk of the prey of the majoriry of secretary birds. The birds often congregate around the edges of bush fires to capture snakes, lizards, small mammals, and many other creatures that flee from the flames.

where along the body as the eagle drops onto it, but the body is quickly passed through the talons until it is held just behind the head. Talons and beak then crush the head and kill the snake. Slender snakes may be lifted up as soon as they have been caught, but they find it difficult to strike under such conditions, even if they are being carried near the middle, so the eagle is safe from even the most poisonous snakes. Larger prey must be killed on the ground, and the eagle relies on its agility and thick plumage to prevent it from getting bitten. The large constricting snakes, which could easily crush an eagle, are not attacked.

Large snakes are dismembered before they are eaten, but others are eaten whole and it is surprising just how large

a snake can be swallowed in this way. They are usually taken down head first and, whereas most birds store their food in the crop for a while, the snake eagles take the snakes straight down through the stomach and into the intestine. A snake that is being carried back to feed a mate or a youngster – never more than one – in the nest is always left with a small part hanging from the beak. Back at the nest, the waiting eagle gradually pulls at the snake's tail and draws it out of the first eagle. When the whole snake has been withdrawn it is swallowed, again head first, by the recipient bird.

The secretary bird
The secretary bird (*Sagittarius serpentarius*) is another well-known snake-killing raptor, although it does not belong to the same family as the eagles. It is a tall, crane-like bird that lives on the African savannas and in bush country. It gets its unusual name from the tuft of black-tipped feathers on the head, which recalls the cluster of pens that 18th century clerks often carried in their wigs.

Secretary birds spend most of their time walking on the ground, and so do not have the grasping talons so prominent in most other birds of prey. Although known as snake-killers, they really feed on a very wide range of animals that they disturb as they walk through the grass, often stamping hard to increase their effectiveness. Small mammals, locusts and other insects, and lizards are the main foods, but small tortoises are snapped up – eleven were found in the crop of one bird – and snakes are killed whenever they are found. The snakes are trampled by the secretary bird's heavy feet until they are almost or quite dead, and then the beak takes over. The wings are often used as shields to protect the body from possible snake strikes.

The Everglades kite
This bird, which is not closely related to the true kite, can be found living in swamps and marshes in South and Central America. It is an unusual bird of prey on two counts: it feeds on invertebrate prey, and it restricts itself to just one kind of food – the apple snail (*Pomacea*), which is abundant in the swamps. The kite is never likely to go short of food as long as the swamps remain, and it is almost certainly the regular supply of this one kind of food that has allowed the bird to develop its special technique. It glides or hovers over the swamp until it sees a snail, and then drops down to pick it up in the feet. It flies back to its favourite perch and holds the shell gently until the snail starts to emerge, and then the bird strikes with its very fine and strongly hooked beak. The snail is speared and, in its death throes, it jerks itself loose from the shell. The bird can then shake off the shell and eat the snail – an interesting contrast to the energetic shell-bashing technique that the song thrush uses.

A diet of snails – every day
Below The Everglades kite feeds only on a certain kind of water snail that it finds in swamps and marshes. Holding the shell in one foot, the bird waits for the snail to emerge and then spears it with the narrow beak.

Owls-Silent Killers of the Night

The owls have been described as cats with wings, for both groups of animals are admirably adapted for catching mice and other small mammals at night. Owls are really the nocturnal equivalents of the hawks and eagles, although they are not quite so diverse in their feeding methods, and there are several marked similarities between the two groups of birds. The hooked beak and the powerful talons are very similar, for example, although the owls generally keep their weapons more or less hidden among their dense plumage. The two groups are unrelated, however, and the similarities are due entirely to convergent evolution (see page 116). A number of owls perform during the daytime, but most of the 130 or so owl species start work at nightfall and their amazing eyesight and hearing allow them to go on hunting successfully right through the darkest night.

An eye for the night

The big, forward-looking eyes are set wide apart on the flat face and they give the owl good stereoscopic vision over a field of about 70°, thus allowing it to judge distances very accurately. But it is the ability to see in what we might consider to be complete darkness that really sets the owls apart from the other birds of prey. No owl can see in absolute darkness, because eyes can function only when light enters them and stimulates the nerve cells of the retina, but it is never completely dark at night: even on a cloudy, moonless night there is always some light from the sky. Illumination, which is the amount of light falling on a given area, is measured in foot candles. This is a complicated unit to define, but some idea of its value can be obtained from the fact that a full moon at its zenith illuminates the earth to the level of about 0.02 foot candles. Even on a cloudy and moonless night, the illumination of the earth's surface is rarely less than 0.0004 foot candles, but experiments have shown that several kinds of owls can see their prey when the illumination level is only 0.00000073 foot candles. This would be the illumination produced by one ordinary candle over 400m (1312ft) away; our own eyes would detect absolutely nothing.

The explanation for the owl's astounding eyesight – probably 100 **times** more sensitive than our own – lies in the size and construction of its eyes. The large eye can obviously gather in more of the available light than a small eye, but the ability to use this extra light to the best effect rests on the structure of the eye. The eyeball of a day-flying bird is more or less spherical, with the retina a long way from the lens. The image thrown onto the retina therefore loses much of its brightness. This does not matter in good daylight, but the night-flying owl cannot afford to lose brightness and its eyeball is shaped quite differently so that the retina is much closer to the lens and thus receives a much brighter image than would otherwise be possible at night.

The owl's retina itself differs from that of day-flying birds in its complement of light-sensitive cells. Day-flying birds have retinas composed largely of sensitive cells called cones. These detect different

colours, but they can work only when the light intensity is above a certain level. The rest of the retina consists of cells called rods. These cannot detect colours, but they work at very low light intensities. Our own eyes contain both rods and cones, but only the rods work at night, so we see everything in black and white. The owl's retina, however, consists almost entirely of densely-packed rods

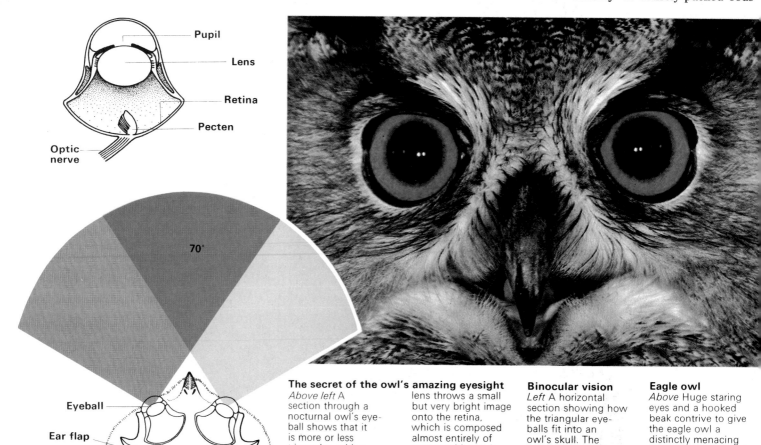

Pupil
Lens
Retina
Pecten
Optic nerve

70°

Eyeball
Ear flap
Skull

The secret of the owl's amazing eyesight
Above left A section through a nocturnal owl's eyeball shows that it is more or less triangular, with a very large pupil to let in the maximum amount of light. The large, rounded lens throws a small but very bright image onto the retina, which is composed almost entirely of very sensitive rods. The feathery pecten is thought to provide nutrients to the inside of the eyeball.

Binocular vision
Left A horizontal section showing how the triangular eyeballs fit into an owl's skull. The overlapping fields of vision of the eyes gives the bird binocular vision over an arc of 70°.

Eagle owl
Above Huge staring eyes and a hooked beak contrive to give the eagle owl a distinctly menacing appearance. The beak does not look very large, but the bulk of it is hidden among the feathers.

Approach path
Right The gliding flight of a tawny owl (*Strix aluco*) homing in on its prey. The head and eyes are fixed on the victim until the very last instant, when the talons are swung forward to make the kill. The broad wings provide good manoeuvrability and also give the lift necessary for the bird to rise with its prey.

Element of surprise
Left The large wings of the tawny owl are able to generate tremendous power with relatively few wing beats, and yet they are almost noiseless in action. This not only enables the owl to hear the scratching and squeaks of its prey but also give it the vital element of surprise when it strikes.

Silenced feathers
Below One of the flight feathers of a tawny owl, showing very clearly the fine fringes along the leading edge and on part of the trailing edge. The wing-tip feathers separate to some extent in flight (as shown in the large colour illustration), and both edges need to be silenced.

Ready to strike
Above A tawny owl about to strike, showing the immense, sharp talons widely spread to cover moving prey. The eyes and ears focus on the victim, as the talons swing forward.

and the bird thus has very limited colour vision but very clear vision in black and white, even when we can see nothing.

A good ear
As well as their incredible eyesight, the owls have very acute hearing, and this plays an important role in detecting prey on the ground. Mice, voles, and shrews, which together make up the bulk of an owl's food, give out a variety of high-pitched squeaks and also make many

small rustling sounds as they move about, and owls' ears are very sensitive to these sounds.

The ears are situated well out to the sides of the head, and when a sound occurs to one side of the bird, the nearest ear picks it up a fraction before the other ear. The time lapse if only about 1/30,000th of a second, but it is enough for the bird to determine from which side the sound came and turn towards it. The eyes might then be brought into play,

but if the sound is more or less continuous the owl can home in on its prey using just its ears. The ears are somewhat asymmetrical and they are arranged so that the bird hears the sound at its loudest when it is coming from directly in front of the head. By orientating itself so that it maintains the maximum volume, the owl can bear down on its victim with an error of no more than one degree either side of the target.

Experiments have shown that the

barn owl can catch its prey in this way in complete darkness, but infrared photography reveals that it does not use the normal approach glide under such conditions: with its head firmly fixed on a collision course with the victim, it flaps down and then brings its talons forward to snatch the prey at the final moment.

On silent wings

Since owls use their ears to track down their prey, it is not surprising to find that the night-flying species have evolved several devices to ensure virtually silent flight. The wings are relatively large compared with the size and weight of the body, and this enables owls to generate a lot of lift with little effort. Owls can glide easily and patrol their hunting grounds with relatively few wing-beats. Even when they do indulge in flapping flight, the feather structure ensures that they are virtually noiseless.

There are three silencing devices associated with the wing feathers. First, the leading edge of the primary feathers has a stiff, comb-like fringe formed by extension of the feather barbs. This fringe is 4mm (0·16in) wide and acts like a series of slots to smooth the air-flow over the upper surface. It also reduces the impact of the leading edge on the air and therefore cuts down noise even further. Secondly, there is a fringe on the trailing edge of the flight feathers. This is formed by some very flexible extensions of the

Asymmetrical ears
Right The head of a tawny owl drawn from the right and left sides with the facial disc of feathers pulled forward to reveal the large ear openings. These are asymmetrically arranged, with one opening higher than the other and with different arrangements of flaps. This asymmetry helps the owl to detect both the direction and distance of a sound source.

Real and false ears
Left The head of a long-eared owl, showing the prominent facial disc of feathers, which is typical of most of the really nocturnal owls. The true ear openings are concealed just behind the disc. The long 'ear tufts' have nothing to do with the real ears; they are simply movable tufts of feathers, and their position may indicate the mood of the owl.

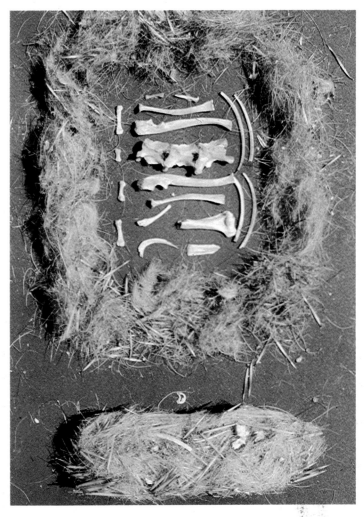

A successful hunt
Left A barn owl returns home with a vole. Although the prey is caught in the talons, it is carried in the beak because the feet are needed for landing.

Evidence of an eagle owl's meal
Above A pellet coughed up by an eagle owl may be over 10cm (4in) long and usually contains the fur, feathers, and bones of a number of prey animals, including rodents, hares, and birds as large as the capercaillie. An assortment of bones extracted from such a pellet are shown at the top of the picture.

barbs that smooth out the mixing of the upper and lower air streams, thus eliminating turbulence and noise. Thirdly, the upper surface of most of the wing feathers is rather downy, and this muffles the noise as the feathers slide over each other in flight and air rushes between them during the up-stroke.

As well as enabling owls to hear the squeaks of their prey, silent flight gives them the element of surprise and allows them to pounce without warning. It is interesting to note here that fishing owls and other day-flying species lack these silencing devices. These owls hunt entirely by sight and, like the hawks and eagles, they are swift fliers and rely principally on the speed of their attack to surprise their prey. It is also interesting to note that the wings of the nocturnal, insect-eating nightjar have silencing devices similar to those of owls.

The hunt
Although owls have specialized in nocturnal hunting, they do not all hunt in the same way. Three main methods are used. Many owls, including the common tawny owl (*Strix aluco*), merely sit on a perch and look and listen until something stirs, and then glide straight down to snatch up the victim. Others, including the barn owl (*Tyto alba*) and the short-eared owl (*Asio flammeus*), swoop silently to and fro just a metre or two above the ground as they search for prey. The short-eared owl is, in fact, the perfect nocturnal equivalent of the hen harrier: both swoop low over the ground and take identical prey but, because one becomes active when the other goes to sleep, they can both live in the same place.

The third hunting method used by owls is much less common and used by only a few species, including the scops owl (*Otus scops*). These birds sometimes use a hawking technique to snatch insects in mid-air, but this form of nocturnal hunting is really the domain of the bats and nightjars. Most insect-eating owls pluck their food from the ground or from vegetation.

The proof of the pudding
Small mammals are the staple food of most owls and, although the skulls may be crushed with a snap of the beak, the animals are usually swallowed whole. Small birds, such as house sparrows, are also eaten by many owls, and in urban areas they can account for as much as 90 percent of the diet. Some of the larger features may be pulled out first, and the small bird may be dismembered before it is eaten. In common with other birds of prey, owls cough up the indigestible remains of their victims in the form of pellets, but owl pellets are usually much more interesting than those of the hawks because owls are unable to digest bones. The pellets therefore contain the skeletons of their prey and it is possible to work out exactly what owls have been eating by examining the contents of the pellet. This can be done when the pellets are dry, but the small bones may get broken as they are pulled from the matted fur. It is better to soak the pellets in water for a while to soften them.

The Fearsome Hunting Wasps

Strictly speaking, almost all wasps are hunters, for they nearly all capture animal food with which to feed their young. The social wasps, or yellow-jackets, that feed on our fruit and jam in the summer are savage killers on behalf of their younger siblings in the nest and they kill a wide variety of other insects. Wings are torn or bitten off, and the bodies of the victims are then chewed up before being fed to the wasp grubs. The term hunting wasp, however, is usually applied only to the solitary species – those in which each female makes her own small nest or nests. There are no large colonies and no workers. The only species that do not hunt are the cuckoo wasps which, as their name suggests, lay their eggs in the nests of other species.

Living larders

The solitary or hunting wasps – nectar-feeders as adults – provide their offspring with flesh, but they go about it in a way quite different from that employed by the social species. In contrast to the social wasps, which bite their victims to death, the solitary species always sting their prey. The sting paralyzes the victims, but it does not kill them and their bodies thus remain fresh for the young wasps to eat. Some of the solitary wasps provide regular supplies of food as their larvae grow, but the majority go in for stock-piling or mass-provisioning: they fill their nests with paralyzed prey sufficient for the full development of the larvae, and then seal up the nests and have nothing more to do with them.

Each of the hundreds of species of hunting wasps, which far outnumber the social species, has its preferred kind of prey. Some take only spiders, while most of them specialize in certain kinds of insects. There are thus cicada-hunters, cricket-hunters, caterpillar-catchers, fly-catchers, bee-killers, and many more.

The initial detection of the prey may be by sight, but scent also plays an important role in determining whether a potential victim is of the right kind. Some caterpillar-catchers rely almost entirely on scent to find their well-concealed prey, and they can be seen running over the vegetation and tapping the leaves with their antennae to pick up any traces of caterpillar scent. Cricket-hunters and cicada-hunters may be attracted by the sounds of their prey, but scent in the air or on the ground is the major clue for these hunters.

The spider-hunters

Most of the spider-hunting wasps belong to the family Pompilidae. These wasps generally attack spiders as large or larger than themselves, and many

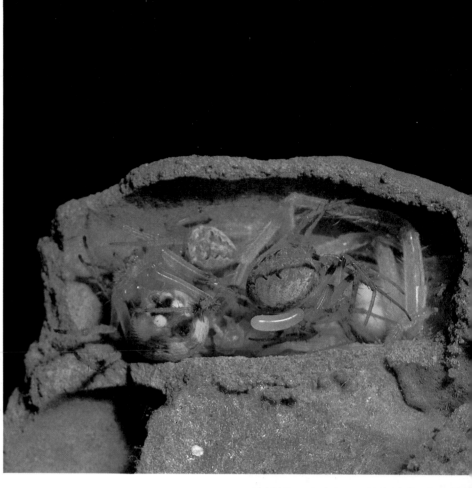

exciting battles have been witnessed between two such adversaries.

The largest spider hunters belong to the genus *Pepsis* and they are known as 'tarantula hawks' in North America because of their attacks on large trap-door spiders and tube-dwelling wolf spiders. The wasps may actually enter the burrows to do battle, but they more often entice the spiders out into the open. In a remarkable display of agility, the wasp then dances around the spider, avoiding the poisonous fangs and eventually managing to plunge its sting into the nerve centre on the underside of the spider's body. A preliminary sting is sometimes administered on the spider's head, serving to put the fangs out of action if not to paralyze the whole spider.

As soon as the spider has been sub-dued, the *Pepsis* wasp sets about getting it back into its burrow, dragging and pushing with surprising strength until the spider is well buried. An egg is then laid on the inert victim and the burrow is sealed and camouflaged with debris. The wasp takes no further interest in the burrow and goes off to find and bury another victim. Each buried spider acts as a living larder, providing just the right amount of food for the development of one wasp larva.

Smaller pompilids attack proportion-ately smaller spiders, but one spider

usually suffices for the development of each larva. Most of these spider-hunters capture a victim and then start to make a nest burrow in the nearest suitable spot, but some dig the burrow first and then go in search of prey. This method involves transporting the prey, some-times over considerable distances, and the wasp may make things easier by biting off the spider's legs to reduce the drag. The wasp may fly short distances with its prey, but generally heaves it along the ground.

It often fails to find the burrow at the first attempt, and may lodge the para-lyzed spider up in a plant while trying

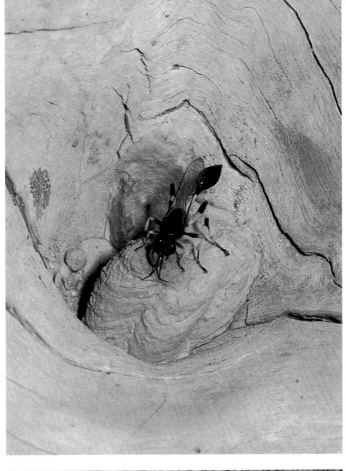

Mud-dauber wasp
Above right Sceliphron
uses her jaws to collect
mud and fashion it into
chambers for her eggs.
When complete, each
chamber is stocked
with paralyzed spiders
(above left) and a wasp
egg. It is then sealed up.

Tarantula – killer
Left Pepsis hunts large
spiders and sometimes
fights them in their
burrows. It nimbly
avoids the poison fangs.

The sand wasp
Right Ammophila hunts
only caterpillars. After
paralyzing them with
several stings, she puts
them in her burrow.

the body to immobilize all the legs. Some species of *Ammophila* collect large caterpillars and drag them back to their burrows, putting just one into each cell. Other species collect smaller larvae and fly back with them gripped firmly in their jaws. They put several caterpillars into each chamber.

When each cell has been provisioned, *Ammophila* lays an egg. In many species the cell is then sealed and that is the end of the proceedings, but some species, such as *A. pubescens*, continue to bring prey to the nest at intervals as the young wasps develop. The wasp may even have several nests on the go at once, and she has the 'intelligence' to know when to deliver new food supplies to each one.

The bee-killers

Wasps of the genus *Philanthus* stock their nests with paralyzed honeybees, which they normally capture on the flowers where the bees are busy gathering nectar and pollen. The wasp detects the bees first by sight, and may even follow a bee in flight. At this stage it is responding purely to a moving image of about the right size, but as the wasp gets closer its sense of smell takes over. Only if it detects the correct scent will it go in for the kill, and this is why *Philanthus* is never deceived into taking drone-flies and other insects that mimic bees. The mimicry is purely visual and its function is to protect the mimics from birds and other predators that hunt by sight alone. But *Philanthus* is not fooled and, if the wrong scent signals are received from an insect, the bee-killer turns away. If the right scent signals are received, the wasp homes in on the feeding bee and takes it by surprise. Keeping well away from the bee's own sting, the wasp curves her body round and stings the bee on the underside of the thorax.

Paralysis follows very quickly and *Philanthus* flies away with the bee slung upside down under her body. She may go straight to the nest, but more often she alights on a convenient leaf and begins to knead the bee's body. Nectar oozes from its mouth, and the wasp eagerly laps this up. It then carries the bee away to the nest, which is usually escavated in a sloping bank. Like all burrowing wasps, *Philanthus* covers the nest entrance with soil before it goes out, but it learns to recognize surrounding objects as landmarks and usually finds the nest again quite easily.

Unlike *Ammophila*, which has to put its booty down before opening the burrow, *Philanthus* can alight on its back legs and use its front legs to scrape the soil away from the entrance, while still holding the prey securely under its body with its middle legs.

to find the way back. Putting the victim up above ground level reduces the chance of its being carried away by ants.

The spider-hunters of the genus *Trypoxylon* belong to a different family and they stock their nests with relatively small spiders. The nests may be excavated in plant stems, or constructed with mud, and usually contain several chambers. Each chamber is crammed with spiders and sealed up with an egg.

As with most solitary wasps, only the female is usually involved with nesting, but in some species of *Trypoxylon* the male lends a hand. He remains close to the female and guards the nest while she

is away collecting food, thus preventing the entry of 'cuckoos' and parasites.

Caterpillar-catchers

Caterpillars are the staple food of a large number of hunting wasp grubs. Female potter wasps make little clay pots or vases and stock them with non-hairy caterpillars. An egg is laid in each pot, either before or after stocking it.

The slender-bodied sand wasps (*Ammophila*) excavate more or less vertical burrows in the ground and construct one or more chambers in each burrow. They then go in search of caterpillars, which they sting at several points along

Pirates and Scavengers

Many birds of prey take food in the form of carrion when they get the chance. Even the magnificent golden eagle does not object to feeding on dead rabbits and sheep, and it may actually take more carrion than live food in the winter. Carrion offers a good meal for virtually no effort, and it is not surprising that some of the birds of prey, notably the vultures, have come to rely almost entirely on this source of food. These scavenging habits are best developed in mountainous regions and in the savanna areas, where large animals live — and die — and where their bodies are easily seen from the air. Vultures are especially common in areas of primitive animal husbandry, where sickly cattle and other animals are allowed to roam until they drop. Predatory mammals are usually kept down in such places, and the vultures get rich pickings.

The vultures — ugly but effective

Vultures belong to two distinct groups, which we can conveniently refer to as New World vultures and Old World vultures. The six New World species belong to the family Cathartidae, while the fifteen Old World species belong to the Accipitridae — the same family as the eagles, from which they are believed to have descended. There are quite a lot of anatomical differences between the two groups, but similarity of habits makes them outwardly very similar.

Most vultures kill antelopes and other small mammals from time to time, but they usually take only sickly animals. With a slow approach flight and relatively weak feet and claws, they are simply not equipped to make an efficient kill. The whole body is designed for a diet of carrion. The beak is hooked and sharp-edged for tearing and cutting skin and flesh, and the tongue is rough for rasping flesh from the bones. The most striking adaptation, however, is the baldness of the head and neck, which allows a vulture to plunge its head deep into bloody carcases without having to contend with blood-soaked feathers afterwards. Blood congealing on the bare skin soon dries and flakes off. Exposure to the sun and air also kills bacteria picked up on the skin.

Large quantities of bacteria are ingested with the decaying meat, but the vultures can eat even the most putrid carrion with impunity because their digestive juices have a powerful antiseptic action and destroy all the bacteria. Even the birds' droppings seem to have antiseptic properties: they are squirted over the legs and feet and presumably keep these organs free from infection.

Vultures find their food entirely by sight. Experiments have shown that they

Scavenging hordes
Above A marabou stork watches a group of Ruppell's griffon vultures tearing at a hartebeeste carcase and waits for an opportunity to join in the feast.

Ever watchful
Right Even on the ground around a carcase, the white-backed vultures remain alert for hyenas that might drive them from their food. This vulture is one of the first to eat, taking mainly offal and other soft meat.

cannot find even the smelliest carcases if they are covered up, but that the birds do come down to investigate cardboard models that have been painted to look like carcases. The usual method of finding food is for the vultures to soar high into the sky on their broad wings and to scan the ground from heights of up to 2000m (6560ft). The birds are often said to have fantastic eyesight, but it is not really as good as that of some of the hawks and eagles. A dead antelope or a pride of lions gathering round a zebra are quite easy to see, even from 2000m, and the vultures certainly do not have the good binocular vision and distance judgement that the hawks and eagles need for swooping straight down onto their prey at high speed.

Vultures coast down leisurely, land near the food, and waddle up to it. It does not take long for quite a group to congregate around a carcase, some of them coming from as much as 60km (37 miles) away. They do not see the food from such distances, of course: each bird soaring in the air watches its neighbours as well as watching the ground, and if one bird goes down its neighbours are likely to follow, bringing even more birds down with them in a kind of 'domino effect.'

Queuing for food

The vulture flocks that congregate around carcases on the plains of Africa are generally mixed flocks, with perhaps six different species in them. They are sometimes lucky enough to find a com-

The Arctic skua – pirate of the skies
Above An Arctic skua chases a lesser black-backed gull through the air in an attempt to make it disgorge the fish that it is carrying back to its youngsters in the nest.

plete carcase for themselves, but they more often have to wait until a pride of lions have finished eating before they can move in, and even then they must keep an eye open for hyenas. Although there may be as many as six species of vultures in the flock, they do not all feed at once; there is a definite hierarchy, based mainly on size, and the birds queue up to take their turn. Each species has slightly different food preferences, and so all the birds get something to eat.

On the African plains the first vulture to eat is often the lappet-faced vulture (*Torgos tracheliotus*), a large and powerful bird but one that eats mainly the skin and offal from a carcase. This is the bird that first rips open a fresh carcase. Next in the queue is usually the white-backed vulture (*Pseudogyps africanus*), the commonest of the East African species. This bird eats some skin and tendons, but it prefers offal and other soft and easily removed pieces of meat. The white-headed vulture (*Trigonoceps occipitalis*) generally comes next, taking large chunks of red meat and a certain amount of skin and gristle. The way is then open for the hooded vulture (*Necrosyrtes monachus*), which has a rather small beak and digs out small pieces of meat from the crevices. All the species mentioned so far are gregarious birds, but the Egyptian vulture (*Neophron percnopterus*), tends to be solitary. A single individual can often be seen around the

flocks of other species, darting in now and then to snatch small morsels. It is particularly adept at scraping small pieces of meat from the bones and it is often the last to leave a carcase.

The relatively handsome lammergeier (*Gypaetus barbatus*) is the final visitor to a carcase in some areas. Unlike other vultures, this species has feathers on its head and neck, and it clearly does not plunge its head into rotting corpses. It gets a good deal of its nourishment from the bones and scraps left behind by the other scavengers. Small bones are swallowed, while larger ones are carried into the air and dropped onto the ground until they break. The lammergeier then uses its long tongue to scoop out the marrow.

The bird lives mainly in mountainous regions, and the chances of dropping the bones onto a sufficiently hard surface are quite high. Small tortoises are dropped in the same way, and the lammergeier makes up its menu with a variety of small animals and scraps. It visits human rubbish dumps to scavenge, and it is even said to lay in future food supplies by knocking goats and other animals from rocky ledges.

The American vultures include the majestic Andean condor (*Vultur gryphus*), which has a wingspan of about 3·5m (11·5ft) and is the largest of all birds of prey. It feeds mainly on the abundant carcases to be found in its mountainous home, but it also kills lambs and young deer. Other New World vultures, which are generally called buzzards in America, include the turkey vulture (*Cathartes aura*) and the American black vulture (*Coragyps atratus*). The latter is extremely common in tropical and subtropical America and does a great deal of its scavenging in towns.

Aerial pirates

Piracy among birds involves the stealing of prey that has been caught by another species. It can be thought of as premature scavenging or 'lazy' predation, although some pirates actually work harder than they might have to if they caught their food themselves. The habit has evolved in several groups of birds, including the birds of prey, the frigate, or man o' war birds (see page 168), and the skuas. The latter are relatives of the gulls and, like the gulls, they spend much time at sea.

The great skua (*Stercorarius skua*) lives in both the Arctic and Antarctic regions, but the other three species are found only in northern waters. Terns and gulls are the main victims in the north. They are chased with great speed and agility and forced to drop or disgorge the food they are carrying. The skuas can pick out the food-carrying birds, even when the food has been swallowed, by detect-

ing their slower flight. Gannets are sometimes attacked by actually bumping into them or grabbing a wing-tip. Knocked off balance in this way, the gannet immediately disgorges its load, and the skua swoops down to catch it in mid air.

Piracy is most common during the breeding season of the victims, when large numbers of birds tend to congregate in one area and constantly ferry food back to their nests. At other times of the year, the skuas catch their own fish or prey directly on smaller birds, and they often scavenge along the seashore.

The great skua, also known as the bonxie, is less piratical than the other species, although it does harry gannets quite often. It catches a lot of fish, and it also preys on other birds, such as kittiwakes, puffins, and various land birds. These are chased through the air and struck with feet and wings, and even head or breast, until they fall exhausted. The skua does not have the sharp talons of the birds of prey to make a clean kill, but its sharply hooked beak is well able to tear open the bodies of its prey. Penguin chicks and eggs are common prey in the Antarctic region. The skuas sometimes pair up to drive penguins from their nests, but more often they merely swoop down to snatch unguarded eggs or sickly chicks. The penguins' chicks habit of clustering tightly together affords them some protection from the marauding skuas.

Dragonflies and Robber-flies

The word dragon is believed to have been derived from an Ancient Greek word meaning 'to see clearly.' The dragonfly is thus very well named, for its eyes are immense – often occupying almost all of the head – and it can detect its small insect prey from a distance of several metres.

There are two distinct groups of dragonflies – the Zygoptera, or damselflies, and the Anisoptera, or true dragonflies. The damselflies are slender-bodied insects with flimsy wings on which they appear to drift from plant to plant. They take most of their prey from the waterside vegetation. The true dragonflies are stouter insects, with broader and tougher wings and a much stronger flight. Almost all their prey is caught in the air. The following paragraphs apply mainly to the Anisoptera.

Thousands of lenses
The dragonfly's huge, bulging eyes, combined with the extreme mobility of its neck, give the insect vision in virtually every direction. This is what makes the dragonflies so difficult to catch with a net, and it also enables them to spot prey anywhere within about 20m (65·5ft). Some dragonfly species may be able to detect moving prey as much as 40m (131ft) away. Each eye may contain up to about 30,000 separate facets, or lenses, each sending a signal to the brain and contributing to a complex mosaic image.

The facets in the upper part of the eye are often larger than those in the lower half, suggesting that the upper ones are primarily involved in detecting movement while the more closely set lower ones can pick out detailed shapes. Recent work in Russia suggests that when the upper facets detect something flying above them, the dragonfly responds by flying up above the object. It may then use its lower facets for the chase. The lower facets are also used in those relatively rare instances in which anisopterans snatch insects from the vegetation. Here it is obviously essential for the predator to be able to pick out the form of its prey, which is often sitting quite still on a leaf or a twig.

Flying gnat traps
The dragonfly uses its legs to capture flying prey, and the whole of its thorax has been modified to this end. The second and third segments have become tilted so that the spiky legs are thrust forward under the head during flight. They form a 'net' that scoops up all kinds of other insects as the dragonfly skims gracefully to and fro through the air.

Anything that will fit into the net will be taken, although very tough insects, such as some beetles, will be rejected. Very often dragonflies can be seen flying close to bee hives and wasp nests, waiting to pick off the bees or wasps as they fly in and out, but the bulk of their prey consists of gnats and other small flies.

The food is transferred to the strongly-toothed jaws as soon as it has been caught, and it is generally eaten in flight, although large insects may be carried to a perch to be dismembered and chewed. Only the wings are discarded as a rule.

Hawkers and darters
Dragonflies are divided into two groups, hawkers and darters, according to the way in which they hunt. The hawkers, which include the longest-bodied species, remain on the wing for long periods, patrolling a definite territory such as a stretch of hedgerow. Darter dragonflies are generally stouter than the hawkers and do their hunting from perches.

The regular to-and-fro flight of the hawkers is punctuated by numerous diversions as the dragonfly sees prey and dashes off to investigate – and usually to catch it. Hawker dragonflies are also known to follow large mammals in grasslands, scooping up the numerous insects that are disturbed by the animals.

Some feeding territories may be over water, but dragonflies hawking up and down over ponds and streams are more likely to be searching for mates. These mating flights are characterized by much hovering – which is rarely, if ever, seen during feeding flights – and by regular clashes when two males meet at the boundaries of their territories.

In temperate regions the feeding flights occur at any time, and may go on all day, but in tropical areas feeding flights are generally restricted to the hours around dawn and dusk. The dragonflies would get too hot if they continued to fly in the heat of the day.

Bare twigs and grass stems are favourite perches for darter dragonflies, but quite a number of species perch on the ground. When a likely victim flies by, the dragonfly darts out to snatch it and eat it. The darter then returns to its perch to wait for the next course. It frequently perches in a horizontal position and uses its front legs to clean its eyes and face.

Darters fly throughout the daytime and, in tropical regions, display interesting methods of regulating their body temperatures. As the ambient temperature rises, many species choose higher perches, away from the hot ground. Others orientate their bodies so that the sun's rays do not fall perpendicularly onto the abdomen; some of the ground-perching species actually sit with their abdomens pointing vertically to the sky. The wings may also be lowered in very hot conditions to act as sunshades to protect the thorax.

Skilled flyers
There has been much discussion about the speeds at which dragonflies fly, especially among entomologists who have tried to catch these nimble insects. Speeds of up to 100kph (63mph) have been claimed, but these are certainly wild exaggerations. It is unlikely that any dragonfly can reach air speeds of even 50kph (31mph), although the insects' aeronautical skills often give the impression of greater speed. Their

Finding prey with compound eyes
Below The damselfly locates prey with its bulging compound eyes. Their range is increased by the elongation of the head.

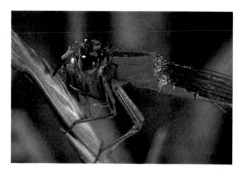

four wings work independently and allow the insects to make incredible manoeuvres – instant stops, 90° turns, hovering, and even backward flight.

This manoeuvrability enables the dragonflies to make efficient use of the mating swarms of gnats and other small insects. The males of these insects gather in great swarms and dance up and down in one place to attract the females. They also attract the dragonflies, which are able to dance up and down in the swarm with them and catch as many as they can eat. This is much more efficient than

Sucking a damsel fly
Above The robber-fly uses its legs to hold its prey whilst its piercing mouthparts (right) are plunged deep into its tissues. As the prey collapses very rapidly after the proboscis of the robber-fly is inserted, the injection of a poisonous toxin is suspected.

Dismembering a bee
Left The legs of the dragonfly are situated well forward on the thorax to hold the prey near the biting mouthparts. The stout mandibles are toothed and used to dismember the corpse.

the swifts and swallows, which have to dive through the swarms time and time again to obtain a meal (see page 154).

Robber-flies

Dragonflies are not the only insects that catch their prey on the wing. The ascalaphids, which are related to the lacewings and ant-lions, employ very similar methods to those of the dragonflies, although they tend to hover much more in the sunshine. Many true flies (order Diptera) also specialize in aerial warfare, and none is better adapted for

this way of life than the various species of robber-flies in the family Asilidae. These nearly all live by chasing or intercepting other flying insects.

Robber-flies are all rather bristly flies, the majority of which are stoutly built and endowed with powerful legs. The large eyes are separated by a deep groove in the middle of the head and it seems likely that they give the flies a fair degree of stereoscopic vision and good judgement of distance. The eyes pick up movements very easily, and the fly gives chase as soon as it notices something

moving, but the chase is pressed home only if the object turns out to be a suitable prey. The eyes pick up more detailed information as the fly gets closer to the object.

If the robber-fly does go in for the kill, it swoops onto the prey and grasps it with the first two pairs of bristly legs, which hang below the body like a shrimping net. The hind legs may also be brought into action, holding the struggling prey in a vice-like grip similar to that of *Bittacus* (page 29).

Some species may also use their hind legs to help in their high-speed manoeuvres; species of *Lagodias*, for example, possess broad fringes on their feet and it has been suggested that these could be used as air brakes.

As soon as prey has been caught, the robber-fly plunges in its sharp, horny proboscis, which in some species is strong enough to penetrate a beetle's tough coat. The prey may struggle for a while, and it seems that the dense 'beard' and other bristles on the robber-fly's face function primarily as a protection for the eyes against the struggling prey. The prey usually collapses quite quickly, however, and it is possible that the proboscis injects some kind of toxic material, although there is not yet any definite proof of this.

The prey may be carried some distance, especially if it is small, but the robber-fly usually alights fairly quickly to get on with its meal. The juices of the prey are rapidly sucked up through the proboscis, and the dry husk is discarded in readiness for the next hunt.

Most robber-flies hunt in the manner of the darter dragonflies, adopting a perch, often on the ground, and darting out to grab passing insects. A few species indulge in regular hunting patrols; they spend a good deal of time sitting on a favourite perch, but they make periodic 'inspections' of the surrounding area to see what can be caught.

The robber-flies are not fussy about their food and they will tackle any insect of an appropriate size, including some larger than themselves. Other flies, moths, grasshoppers, ichneumon flies, dragonflies, beetles, and even bees and wasps are caught quite regularly, the stings of the bees and wasps deterring the robber-flies not at all.

Quite a number of robber-flies actually resemble the bees and wasps very closely, but the significance of such mimicry is not always understood. It may help the flies to approach the bees and wasps unnoticed, but they do not prey exclusively on these insects. It seems more likely that the resemblance gives the robber-flies a reasonable degree of protection against birds.

Acrobatic Insect-Eaters

The insects are the most abundant animals on the earth and they provide food for a vast number of other creatures, including many lizards, mammals, and amphibians. Fishes eat large numbers of young insects in the water, and predatory insects themselves take a heavy toll of their brethren. But the birds are the most important insect-eaters. A pair of blue tits, for example, may deliver 10,000 caterpillars to the nest during the 15 days or so that the youngsters are being fed there. This amounts to one caterpillar every 80 seconds during the hours of daylight, but the birds have little effect on the total insect population.

Most of the insects, like the caterpillars just mentioned, are caught on the ground or on vegetation, but some birds have evolved ways of catching insects in mid-air. Many insects spend a high proportion of their adult life on the wing, and any bird that can make use of this source of food is clearly on to a good thing. Like the dragonflies (page 152), the birds employ two major methods of hunting: they either remain airborne for long periods and hawk up and down in search of food, or they adopt perches and streak out to snatch any insect that comes within range.

Swallows, martins and swifts

The most familiar of the hawkers are the swallows, which are found in nearly all parts of the world. The common swallow (*Hirundo rustica*), known as the barn swallow in America, breeds throughout Europe and much of Asia and North America in the summer, and spends the rest of the year in various parts of the southern hemisphere. During the breeding season it favours villages and farmland, where there are plenty of insects. It can also be seen feeding over parks and inland waters, swooping gracefully through the air and using its long, forked tail as a rudder. The swallow has a 'ceiling' at about 150m (492ft), but actually does most of its feeding at much lower levels over open farmland and water. Its beak is very short, but it has a wide gape, well suited to scooping up insects as it flies along.

It was once thought that the birds merely trawled for food, catching whatever insects entered the beak as it carved a path through the air. This probably does happen when the birds feed on dense swarms of gnats over the water, but most of the insects are caught individually with rapid changes of direction. This is easily seen on farmland, where the birds dart cleverly around cattle and catch insects the mammals disturb.

The swallow is less common in many parts of Europe than it was 50 years ago, mainly through changes in farming

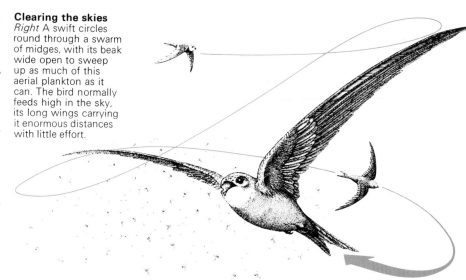

Clearing the skies
Right A swift circles round through a swarm of midges, with its beak wide open to sweep up as much of this aerial plankton as it can. The bird normally feeds high in the sky, its long wings carrying it enormous distances with little effort.

patterns. The traditional farmyard with its muck-heap and its animals is fast disappearing, and with it go a lot of the swallow's food and nesting sites.

The house martin (*Delichon urbica*) is a close relative of the swallow. It started out as a cliff-nester, but found buildings just as suitable and it is now very commonly associated with towns and villages. The martin feeds by snatching insects from the air in the same way as the swallow, but it flies in a much more leisurely fashion and often cruises round at heights of 250m (820ft) or more. This is much higher than the swallow usually goes, and so the two species avoid competing for food. The house martin does swoop low over water, however, and it occasionally feeds on the ground.

It has been suggested that ground-feeding is learned behaviour, acquired by certain populations that find useful amounts of food while collecting mud to make their nests. The normal food of the martin, like that of the swallow, consists of small flies, together with a few small beetles and some of the numerous little spiders that are blown into the air on their silken threads.

House martins are much more common than swallows in towns, where they make use of warm upcurrents that sweep food skywards. The reduction of air-pollution in recent years has led to an increase in insect life and in the house martin population in many British towns.

The swift (*Apus apus*) feeds in the same way as the swallows and martins, although it is not related to them. It generally feeds at even higher levels than the martins – above 300m (984ft) – and is rather more independent of the terrain; it is equally commonly seen over woodlands, towns, villages, farmland, and moorland. Hunting from dawn until long after dusk, it is thought to cover more than 800km (500 miles) each day. Only when the weather is bad and the aerial insects are in short supply does the

Pied flycatchers at their nest
Above Flycatchers do not depend entirely on flying insects. The male pied flycatcher (right) has caught a juicy caterpillar.

swift come down to feed at low levels. Lakes are favourite feeding sites at such times, and rubbish dumps also provide interesting fare. Being a larger bird than the swallows and martins, the swift can deal with relatively large moths as well as flies. It also eats honeybee drones.

The nightjars

The nightjars, typified by the European nightjar (*Caprimulgus europaeus*) and the American whippoorwill (*C. vociferus*), are nocturnal insect hunters living mainly in woodlands. They use the hawking technique of the swallows and, when swarms of midges and mosquitoes are particularly abundant, they scoop them up *en masse* in the wide-open beak. The stiff bristles around the mouth help to net even more insects and funnel them into the mouth, and one nightjar was found to have 500 mosquitoes in its stomach. Moths and beetles are more important dietary items, however, and these are caught individually as the bird

brilliantly coloured, and they all have very similar habits – so similar, in fact, that it is rare to find more than one species in a given habitat. They all catch insects in flight, sometimes while circling around like martins but more frequently by darting out from a perch like the fly-catchers. Carmine bee-eaters (*Merops nubicus*) even have moving perches on occasion: they ride on the backs of bustards in parts of Africa and swoop off to catch the insects that are disturbed by these large birds as they walk through the grass.

Although bees and wasps are the main food of the bee-eaters, they do not ignore other insects; dragonflies and butterflies are commonly caught and eaten, and the birds also take large numbers of locusts. The long beak closes firmly around the prey, often with a loud snapping sound, and the bird may swallow its victim in mid-air or return to the perch with it. Bees and wasps are sometimes battered against the perch, and some observations indicate that the birds may rub out the sting, but it seems likely that the bee-eaters are immune to the venom of these insects and that they swallow them without any discomfort.

Migration

Birds that depend upon flying insects for the bulk of their food clearly cannot survive all the year round in cool climates. Those that breed in Europe and the cooler parts of North America generally migrate to tropical regions at the end of the summer and, although they may have to change their diets to some extent, they are able to find plenty of flying food. Swallows and other species that breed in Europe often spend the winter as far away as South Africa; species breeding in North America generally winter in the Caribbean region or as far south as Argentina; while some birds breeding in northeast Asia may even travel as far as Australia.

One insect-eater that does not migrate, however, is the American whippoorwill. In common with one or two of its relatives, this bird actually goes into hibernation – a phenomenon virtually unknown in the bird world, although it is common enough in the insect-eating bats and some other mammals.

twists and turns through the air, using its large eyes to search for them. Even small birds have been found in the stomach of some nightjars.

A big beakful for the bee-eater
Above Extreme agility is needed to catch a dragonfly, but the white-fronted bee-eater has no problem. It can now enjoy its meal.

Fly-catching from a perch

A number of birds, including the red-starts and the wagtails, fly out from a perch or from the ground to catch at least some of their food on the wing, but few rely on this method of feeding as much as the flycatchers. These are an Old World family of birds with over 100 species, of which one of the best known is the spotted flycatcher (*Muscicapa striata*). This inconscpicuous little bird adopts a territory with a reasonable amount of shrubby vegetation, which encourages insect life, and selects a lookout post within about 2m (6·5ft) of the ground. When an insect comes into view, the flycatcher streaks out towards it, swerving deftly if the insect changes course and then often hovering momentarily before darting in for the kill. While hovering, the bird may be 'deciding' on the suitability of the prey and taking note of possible warning colours. Wasps, for example, are not caught, although bees,

with less prominent patterns are sometimes eaten. Flies are the main prey and the flycatcher's fairly large beak makes an audible snap as it closes round the victim. Even if the aim is not perfect, the stout bristles around the beak help to gather in the prey.

The bee-eaters

The bee-eaters are another Old World family of birds, centred in tropical Africa. Some species are found in southern Europe, and others extend through southern Asia to Australia. Most are

Perching and darting
Right The little bee-eater displays the typical perching and darting behaviour of the bee-eaters and flycatchers It springs from its perch and streaks after a bee, catches it in its long beak, and returns swiftly to its perch. This bird has swallowed its prey in flight.

The Falcons - Feathered Dynamite

The peregrine's stoop

Many of the smaller birds are caught in level flight with the peregrine chasing at speeds of up to 80kph (50mph), but the famous stoop is the usual method of approach, with the peregrine diving almost vertically onto its prey at tremendous speed. During the later stages of the dive, the wings are folded right back and the tail feathers are closed up to reduce the wind resistance to the lowest possible level. The speed of the dive has given rise to much heated discussion, and many widely varying estimates of the maximum speed have been published, but it is generally accepted that a stooping peregrine can reach speeds of about 250kph (156mph). Special baffles in the nasal passages deflect and slow down the air-flow so that it does not damage the lungs as the bird breathes in.

A peregrine's stoop
Above and left Spotting a bird flying some way below, the peregrine folds its wings and plummets groundwards at a speed of up to 250kph (156mph). Just before the strike, the falcon levels out and extends its taloned feet. The lethal, dagger-like claw on its hind toe is driven deeply into the prey, which usually dies instantly. The victim often falls to the ground, but on other occasions the peregrine maintains its hold and carries the prey straight to a favourite perch to be eaten.

The falcons (family Falconidae) are a group of fast-flying birds of prey with long, pointed wings and longish tails. They hunt in a variety of ways, but the typical method is to chase and snatch others birds in mid-air. Only these air-to-air combat specialists are dealt with here. The kestrel and other falcons that catch most of their prey on the ground are described on pages 138 and 139.

The noble peregrine

The best known of the aerial killers is undoubtedly the peregrine falcon, a powerful, crow-sized bird that has been described as a bundle of feathered dynamite. It has long been the favourite among falconers because of its speed and grace, and it was held in such esteem that it was once reserved for use only by the nobility.

The peregrine is one of the world's most widespread and adaptable birds, being found on all continents apart from Antarctica and also on many oceanic islands. It is at home in tropical forest and Arctic tundra, on sea coasts, in the semi-desert, and also high up in the mountains. In some parts of the world it even thrives in the middle of cities; peregrines have been reported nesting in New York, Montreal and Nairobi.

Over 99 percent of the peregrine's food consists of birds, the rest being made up by an occasional insect or small mammal. At least 145 species of birds have been recorded as prey of peregrines in Europe, with pigeons, grouse, blackbirds, fieldfares, and various waders being the most commonly captured species. Ducks, rooks, gulls, auks, and many smaller birds such as larks and pipits are also common victims in the appropriate habitats, but the peregrines show a marked preference for prey weighing between one eighth and one half of their own body weight. With adult male peregrines averaging about 650gm (1·4lb), and females about 1kg (2·2lb), this means that the preferred weight of prey is in the range 80–500gm (2·8–17·6oz). A kill of between 200 and 250gm (8·8oz) each day will meet the needs of an average pair, but more is needed during the breeding season when young are being reared.

Aerial combatant
Top The handsome peregrine falcon stands proudly over its latest victim—a pigeon, which it had started to pluck to get at the rich meat. Pigeons are among the peregrine's commonest prey but, because of their high-speed aerial hunting techniques, these falcons can kill birds much larger then themselves without risk.

The speed of the stooping peregrine gives it the element of surprise, and also enables it to make a clean kill in most instances. The kill is made with the talons, the hind claws striking home so powerfully that the victim is sometimes decapitated in mid-air. Decapitated or not, the prey usually falls straight to the ground and the peregrine swoops after it.

In the rare instances in which the prey has not been killed outright, a quick bite behind the head will finish the job. The falcon then uses its sharply hooked beak to pluck out many of the victim's feathers and to carve the flesh from its body. The wings may be left. but very little else is wasted by the peregrine. Occasionally the stoop ends with the falcon clinging firmly to its prey, in which case it either lands a little further on or swoops up to a convenient perch to enjoy its meal.

It is estimated that about two thirds of the peregrine's serious chases are successful, the failures arising when the prey lands on the ground or joins up with other members of its kind to form a dense flock. The peregrine is loathe to attack a flock of birds, for it could well damage itself by plunging into them at high speed. Likewise, it would injure itself if it tried to dive onto a bird on the ground. Perhaps only one kill in a thousand is made on the ground, and that is when a low-flying peregrine scoops up a small mammal or an insect, or perhaps snatches a small wader from the water's edge.

Over to you

During the breeding season the male peregrine, known as the tiercel, does all the hunting, but he does not usually take any food directly to the nest. He calls to the female as he approaches the nest, and he may then leave the food for her on a nearby ledge. On many occasions, however, there is an aerial hand-over. The female flies out to meet her mate, and may roll over underneath him to take the food in her talons. Alternatively, the male may drop the food when the female is approaching him and she then darts down to snatch it acrobatically in mid-air.

The merlin

The thrush-sized merlin is one of the smallest of the falcons, and certainly the smallest European member of the group. It breeds during the summer on upland moors in Europe and North America, and spends the winter at lower altitudes, often on coastal salt marshes. Merlins eat a few mice and voles, but their main food consists of small birds. These include larks, pipits, stonechats, and various small waders, almost all of which are

caught on the wing. It is rare for a merlin to take anything weighing more than about 100gm (3·5oz), and most of its prey will weigh less than 50gm (1·75oz).

Merlins sometimes indulge in spectacular stoops like the peregrine, but they usually hunt close to the ground like harriers (see page 139). They skim to and fro over the ground until a small bird takes fright and launches itself into the air. Then the chase is on. The merlin zooms after its quarry with incredible agility and persistence, but it is not always successful.

If it spots a potential victim from a distance, the merlin may use a hedge or other available cover to approach unseen until it is near enough to dart out and pounce on the unfortunate bird. The prey is held down by one foot and killed by a deep bite in the neck.

The insect-eating hobby

The hobby is a somewhat larger falcon than the merlin, and yet it obtains a fair proportion of its food from insects, especially during the non-breeding season, which it spends in Africa. The swarms of winged termites that emerge from their mounds in the rainy season are particularly attractive to the hobbies, which sweep through the swarms on their slender wings and gorge themselves for hours on end.

Hobbies breed on heathlands and in open woodlands in Europe, where they feed on a variety of small birds as well as on dragonflies and other insects. The male feeds the female during courtship as well as later during the breeding season, and it is clearly more efficient to catch birds for this purpose than to chase innumerable insects. More than half of the birds caught are larks and pipits, but the rest consist of a very wide range of species. These are usually caught in full flight, with the hobby gliding and swooping gracefully through the air like a large swift. The prey is snatched up by the talons and smoothly transferred to the beak in flight. Unless it is being taken back to the nest, it may be eaten in flight. Hobbies have also been known to snatch small rodents from the talons of flying kestrels.

Prey brought back to the nest is often completely plucked by the male. This is more likely during the early stages of nesting, when he does not have to bring more than two or three birds each day. As the eggs hatch and the fledglings grow, however, the male must bring more prey for them and he does not have time to remove the feathers. This chore is then taken over by the hen bird before she passes the food to the nestlings.

The agile merlin
Left A female merlin tucks into a small bird that she has caught on the wing. Merlins are among the smallest of the falcons.

A family of hobbies
Above Adult and young hobbies at the nest. The youngsters are fed mainly on bird prey, which is always plucked before they get it.

Low-level attack
Left and above The merlin approaches its prey with a low, swift flight, making use of any available cover to hide it from the prey until the last moment.

The Hawks – Agile Woodland Predators

The falcons are primarily birds of open country; in the woodlands, their place as bird-killers is filled largely by the hawks of the genus *Accipiter*. These birds of prey belong to the same family as the eagles and harriers (see page 139), but they are distinguished by their rather short, broad wings and long tail. These features give the hawks their superb manoeuvrability as they fly through the woodlands. The tail acts as a rudder and the birds can turn almost at 90° when they are chasing their prey between the branches.

Sparrowhawks, Ma'am

This is the alleged reply of the Duke of Wellington when asked by Queen Victoria how the sparrows could be removed from some trees that were being enclosed by the Crystal Palace – centrepiece of the Great Exhibition in 1851. We do not know if the hawks were actually used, but they would certainly have been able to do the job. They are highly efficient predators of small birds and, true to their name, they include large numbers of sparrows in their diet.

The sparrowhawk in question is *Accipiter nisus*, which lives in Europe, north Africa, and much of northern Asia. It breeds in fairly dense woodland, but can often be seen hunting in more open situations outside the breeding season. This is especially true of the female. Some small mammals – mainly voles, mice, and young rabbits – are eaten, but Dr Ian Newton's detailed studies in Scotland have shown that almost 98 percent of the sparrowhawk's catches are birds; nine tenths of which were species such as sparrows, finches, pipits, larks, thrushes, robins, starlings, warblers, and tits, weighing on average 100gm (3·5oz) or less. The studies of Dr L. Tinbergen in Holland revealed that 82 percent of the catches weighed less than 50gm (1·75oz).

There are marked differences in the prey captured by the two sexes, but this is not surprising when the male weighs about 150gm (5·25oz) and the female female averages 300gm (10·5oz). The males concentrate on sparrow-sized prey, but they do kill some thrushes and occasionally catch birds as big as themselves. The females are quite powerful birds for their size and they concentrate on the larger end of the prey spectrum. Thrush-sized birds make up the majority of their kills, but they also catch jays and young crows and sometimes kill wood pigeons twice their own size. Newton found that wood pigeons and the crow family accounted for about 7 percent of sparrowhawk kills in his study area and represented about 38 percent of the weight caught. It is estimated that a pair of sparrowhawks need to kill the equivalent of about 1,600 sparrows per year to sustain them-

The sparrowhawk
Above A sparrowhawk glides around a traditional starling roost, waiting for the starlings to return for the night. It attacks them as they settle down on the branches. *Left* A hen sparrowhawk feeds her youngsters. The male brings most of the food to the nest – perhaps eight small birds a day – but never feeds the young.

The largest hawk
Right The goshawk, largest of all the hawks, breeds in the extensive northern forests and feeds on a variety of birds and mammals, such as the squirrel it has here.

selves, allowing for the fact that the female bird needs more food than the male, and that they need a further 400 sparrows or their equivalent to feed an average brood for the eight to ten weeks before the young become independent.

In wooded regions the sparrowhawk generally hunts from a perch, waiting for something interesting to fly into view before it gives chase. It will, however, go out and look for prey if really hungry. When actually giving chase, it flies quite close to the ground, using long glides punctuated by bursts of rapid wing beats that lift it up again. Speeds of 40kph (25mph) are often reached, and some people maintain that the sparrowhawk can achieve even greater speeds.

Whatever its speed, it weaves its way between trunks and branches with amazing precision. In more open situations the sparrowhawk may circle at about 30m (98ft) above the ground while searching for prey and then go into a low-level chase, often making use of hedges and other cover rather like a merlin. Hedgerows are, in fact, the bird's most common hunting grounds outside the woodland habitat.

Prey is sometimes snatched from the ground, but the vast majority of kills are made in mid-air. The long legs, with long, mobile toes and needle-sharp talons, are ideal for snatching fast-moving prey at almost any angle. The sparrowhawk frequently turns on its side or even upside down to capture a swerving bird.

Sparrowhawks tend to take whatever small bird is available, the house sparrow being the commonest victim in many populated regions while the chaffinch may replace it in more remote areas. The proportions of different species also vary with the season: more woodland birds are taken during the breeding season than at other times. But it is not necessarily the commonest birds that are taken in greatest numbers. As with most other predators, ease of capture is an important factor. Swallows may be abundant around the woodland edges, but not many are caught because of their speed. Coal tits are often very numerous in overgrown woods, but they can conceal themselves very easily. Flocking species, such as the sparrows and finches, are attacked so frequently because they are conspicuous. The sparrowhawk chases

a whole flock and, like the hunting dog (page 134) and various other mammalian predators, it is able to detect and concentrate on one of the weaker individuals.

Having made a kill, the sparrowhawk generally carries it to a favourite plucking post, which may be in a tree, on the ground, or literally on a post. The head of the victim is often bitten off and crushed so that the brain can be eaten, and then the sparrowhawk plucks a good deal of the feathers. Large feathers are yanked out by grasping them near the base of the vane, and the feathers often show tell-tale beak marks. The breast muscles are usually the first parts to be eaten after the head, and with relatively large prey these may be the only parts that are consumed by the sparrowhawk.

During the breeding season the sparrowhawk usually has several plucking posts scattered around the nest area, but not near enough to give away its position. The prey is then carried ready-plucked to the nest, and the uneaten bones are taken some distance away before they are dropped. Plucking is not complete, however, and the sparrowhawk's pellets consist largely of feathers,

claws, beaks, and small pieces of bone.

The goshawk

With a length of about 60cm (24in) and a weight in excess of 1kg (2·2lb), the goshawk (*Accipiter gentilis*) is the largest of the hawks. It inhabits large areas of Canada and northern Eurasia. Like the sparrowhawk, the goshawk can sometimes be seen in open country, but it requires extensive forests for breeding and it is most likely to be seen around the forest edges and in clearings. Both coniferous and deciduous forests can support the bird, which is a magnificent sight as it streaks through the trees in search of prey.

Its hunting methods are identical to those of the sparrowhawk, but it can obviously deal with larger prey. Its feet and legs are much more powerful than those of the sparrowhawk and it can overpower mammals considerably larger than itself. Rabbits and hares weighing up to 1·5kg (3·3lb) are often caught, and the bird has even been known to kill cats — wild and domestic — and foxes, but about 90 percent of its food comes from birds. Thrushes and starlings are caught

in large numbers, but the bulk of the goshawk's food comes from the much larger crows and wood pigeons. Game birds are a major source of food in Finland, although they play an insignificant role in the diet elsewhere, and one Danish pair of goshawks hunted little else but black-headed gulls — good examples of acquired tastes or habits in small sections of a population.

The female goshawk is larger than the male, but not markedly so and there is no significant difference in the kills of the two sexes. One wood pigeon will provide more than enough food for two days, and it is estimated that 300 fully grown wood pigeons or their equivalent would feed a pair for a year and also enable them to bring up three youngsters. As with the sparrowhawk, the male does all of the hunting during incubation of the eggs and the early lives of the chicks. The male sparrowhawk may bring food to the nest, but the male goshawk never does until the chicks are about a month old. He calls to the female from a nearby branch and either throws the food to her or leaves it on the branch for her to collect when she is ready.

Bats-The Airborne Killers

Apart from birds, bats are the only vertebrates capable of sustained flight. Although some squirrels and marsupials can glide, they cannot stay aloft for long and must alight to feed. Even among the birds, there are relatively few species that feed mainly or wholly in flight like the majority of bats.

Bats form the second largest order (Chiroptera) of mammals, and number about 850 species. Most of these live in the tropics and only three families (Vespertilionidae, Rhinolophidae and a few of the Mollossidae) penetrate the world's temperate regions. In the tropics there is an abundance of food for most of the year, but in deserts or cooler regions there may be a seasonal shortage of food. Many bats avoid this problem by migrating elsewhere, just as many birds do, but some stay put and hibernate – a curious strategem for a group possessing such mobility.

The majority of bats feed on insects, like the primitive mammal groups from which the bats are thought to have evolved. This means intense competition for food, with a clear advantage gained by any species that develops the necessary adaptations to tackle a new food source. This is why about 200 species have become fruit and nectar feeders. It is easy to see how this might have evolved from an insectivorous diet: a bat attracted to the many insects that live in and on ripe or rotting fruit could soon begin to supplement its diet by licking up the fruit juices. Later it might begin to rely more on the fruit than the insects. It might also become attracted by the sweet scents of flowers and lap up pure nectar instead of rotting fruit. Nectar is energy rich, but lacks protein; a deficiency made up by consuming pollen. A number of bats do this, and have a specially adapted tongue for the purpose.

Most bats navigate by using echolocation to detect objects in the dark. They are thus able to fly at night, in gloomy forests and even in caves where there is no light at all. The same mechanism also allows them to detect their prey. High pitched squeaks, emitted through the mouth or nostrils, echo back from the target and are somehow interpreted by the bat to establish range, direction, size and probably texture too.

Vampire bats

The three genera of vampires (*Desmodus*, *Diaemus* and *Diphylla*) probably represent the most extreme form of dietary specialization among bats. They feed on blood.

Vampire bats are quite small animals only about 10cm (4in) long; much less fearsome than their reputation. They are found only in the warmer parts of central and south America; the ones that make

The scalpel-like teeth of the vampire
Above The short snout of *Desmodus rotundus* contains an array of needle-sharp teeth, which can slice through the skin of a sleeping animal without even waking it up.

spine-chilling appearances in horror movies are usually other species or made of cardboard! The common vampire (*Desmodus rotundus*) is the most abundant and widespread species; the other two species (*Diaemus youngi* and *Diphylla ecaudata*) are much rarer and are poorly known.

It is difficult to see how blood-feeding evolved in bats; presumably the ancestral victims must have been the pigs, tapirs and large rodents that live in the forests of tropical America. The arrival of Spanish and Portuguese conquerors in the fifteenth and sixteenth centuries must have been good news for vampires, because the Conquistadores imported horses, cattle and chickens to the New World for the first time. Cows contain far more blood than tapirs and have become far more numerous, so the vampire populations must have greatly increased in the past 300 years or so. Both *Diaemus* and *Diphylla* seem to have become partial to the blood of chickens.

In search of blood

Desmodus is a shy animal. It is active on dark nights, avoiding moonlight if possi-

ble. It flies very low, only perhaps 30cm (12in) off the ground, until a sleeping victim is located. Lights and mosquito netting normally keep them away from people, but cows tethered in the open or in crude byres are easy prey. The vampire alights and scurries forwards, using its folded wings as forelegs. A vampire can run fast, like a huge spider, and can leap into the air if alarmed. It climbs onto its victim, searching for the warm areas of skin where blood vessels are close to the surface, Then, a quick scooping bite with the needle-sharp incisors opens a wound, usually without waking the host. Vampires will return night after night to attack a host; often a dozen or more will congregate to exploit the same victim.

The vampire does not suck blood, but laps it up, drooling copious quantities of saliva into the wound as it does so. The saliva contains a substance that retards blood clotting, so the wound continues to bleed freely for longer than normal. The vampire does not need to chew its food, so it has only 20 teeth, fewer than any other bat. Each vampire drinks about 30ml (equivalent to a large eggcupful) in less than half an hour and may have difficulty flying with the extra load. The blood is mainly water, so the vampire's kidneys soon get to work and eliminate water from the body; sometimes the vampires have already begun to urinate

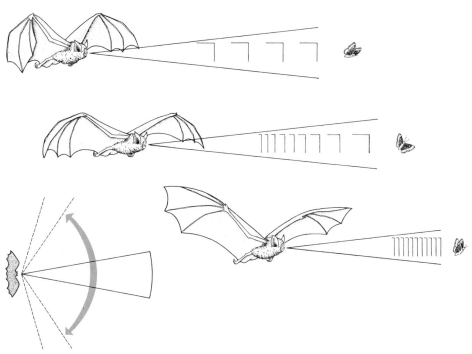

The aptly-named greater horseshoe bat
Above This bat emits its echolocating sounds through its nose. The horseshoe-shaped flap of skin channels them into a narrow beam.

before finishing their meal. The short intestine digests the more nutritious components of the blood and produces black, tar-like faeces that add a characteristic appearance and odour to the vampire's roost.

Vampires cause considerable blood loss; a small colony may consume the equivalent of two dozen cows in a year, quite apart from the flow that continues after they have fed. They also transmit rabies, particularly to cattle, and cause serious losses to South American farmers.

Insectivorous bats
Nearly three quarters of all bats are insectivorous and they catch their prey whilst airborne. It is likely that they can hear the buzz and whine of insects in flight and this will help locate food in the dark. However, it is also true that many bats specialize in taking large fat-bodied insects such as moths and beetles, which are individually tracked by echolocation before capture. Some moths have learnt to take avoiding action when they hear a bat's echolocation sounds, a few even emit confusing 'jamming' sounds of their own to throw the bat off course. Large moths are carried to a convenient perch to be dismembered and eaten. A pile of wing debris may soon accumulate below a feeding roost. A study of British bat prey revealed that over 98 percent of the moths taken were cryptically coloured species, the very ones that escape the notice of birds during the day. However, at night their colours are no protection from an echolocating bat.

Very small insects are probably not followed individually by echolocation because they are too tiny to give adequate echoes. In such cases the bats may

Sweeping the beam
Above The horseshoe bat increases the effectiveness of its narrow beam of sound pulses by sweeping its head from side to side.

Hipposideros sp
Below This bat shows the sharp teeth typical of most insect-eaters.

Sonargrams – sound pictures of the bat's hunting technique
Above These diagrams show three stages in a quest for prey by the greater horseshoe bat (*Rhinolophus ferrumequinum*). The top one illustrates the search, with the bat flying more or less at random and emitting ten pulses of sound per second. Each pulse lasts for about 50 milliseconds (1/20th of a second). The second drawing shows how the pulse rate speeds up when the bat has detected an insect and has started to approach it. Finally as the bat closes in on its prey, the pulses speed up even more, until they come at 75 per second and each lasts for just 10 milliseconds. The speeding up of the pulses in the final capture run ensures that the bat keeps track of every small deviation by the insect.

become 'aerial plankton' feeders and swoop through swarms grabbing what they can; no bat feeds wholly this way.

Insects as small as fruit flies (barely bigger than a pinhead) may be detected and followed by echolocation. As the bat closes in on its prey, it increases the rate of echolocation pulse emissions to fix the target's position more precisely, then grabs the prey with its teeth. Sometimes

a wing is used to brush an insect towards the mouth and occasionally the tail membrane is used to scoop up prey; all this taking place in split seconds while the bat is dizzily gyrating about the sky. The little brown bat (*Myotis lucifugus*) can catch perhaps 500 insects an hour and a pipistrelle may obtain half its own body weight of prey in a similar time.

Certain huge colonies of guano bats

(*Tadarida brasiliensis*) in America must kill over 6,000 tonnes of insects in a year, the individual bats flying at high speed for perhaps 40 or 50km (31 miles) on each of their nightly excursions.

Pouncers and foliage-gleaners
Some individual bats have a regular 'beat', others fly only at tree top height or near the ground. Some fly fast and straight, others with dizzy spirals or weak fluttering wingbeats. Big bats may take prey too large for small bats and not bother with the tiny insects caught by the latter. However, hunting similar prey, by similar methods, bats must compete with each other for food to some extent. Hence the advantage of finding a new strategem. The pallid bat (*Antrozous pallidus*), an inhabitant of the North American deserts, has developed the art of flying low to echolocate large beetles on the ground. It pounces on these, attacking with its teeth, and even somehow manages to catch scorpions by

The rat-catching bat
Left Vampyrum spectrum, America's largest bat, is a real carnivore that often swoops down to pluck rats and other rodents from the ground. It also catches birds on the ground and in the air, as well as other bats.

The long-eared bat
Above The long-eared bat, seen here in full flight, can snatch insects from leaves as well as catching them in mid-air. It emits very quiet sounds when hunting on plants and needs its huge ears to pick up the echoes.

the same tactic – evidently escaping the obvious penalty! Apparently the extra size and food value of fat beetles compensates for the inconvenience to the bat of fluttering on the ground. A quite unrelated species, the slit-faced bat (*Nycteris thebaica*), has evolved similar habits in the dry parts of Africa. This species also haunts latrine pits, presumably in search of a cool roosting place, but also perhaps to find flies and spiders.

The long-eared bats (*Plecotus* sp) have specialized in picking static prey off leaves. This is an amazing feat; not only must the bat hover, a difficult feat in itself, but it must also somehow pick out

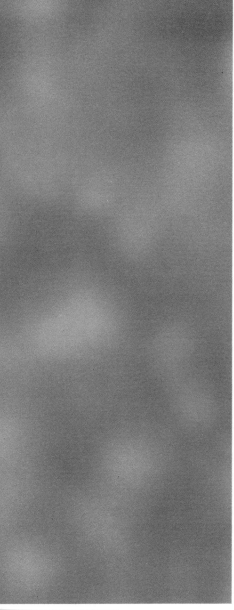

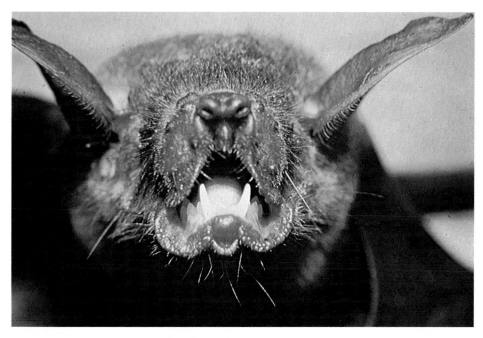

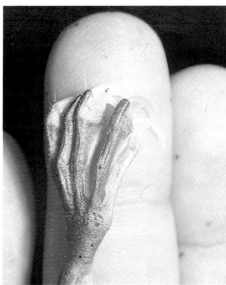

Tools of the trade for the fisherman bat
Left Fishes are very slippery, but America's fisherman bat is well able to cope with them by virtue of its unusually long, hooked claws. These are dragged across the water surface when prey is detected.

Above The fisherman bat's large teeth easily slice up the fish when it has been caught. The bat is clothed with rather short, greasy fur that repels water and prevents the bat from getting waterlogged.

the prey from its background. Many leaf-dwelling insects are hard to see in daylight, let alone in the dark, so the bat must find them by echolocation. Imagine trying to tell the difference between the echo of a caterpillar and that given by the leaf on which it is sitting! Moreover, this must be done at very short range, so the echoes return to the ear less than a thousandth of a second after the sound has been emitted by the bat. To avoid swamping the echoes with noise, long-eared bats emit very quiet echolocation pulses. This is why they need such huge ears to detect the echoes. Though exceedingly difficult, this feeding strategem does enable *Plecotus* to prey on spiders, caterpillars and other non-flying species not available to other, more conventional, species of bats.

Carnivorous bats

Insectivorous bats are carnivores in the sense that they take live animal prey, but a few bats specialize in killing larger

vertebrates for food and perhaps better qualify for the title of 'killers.' The false vampire bats, such as *Macroderma gigas*, can land on the ground and take off easily and use this ability to catch mice, frogs and small birds. They will even swoop into houses to grab geckoes off the walls. The largest American bat, *Vampyrum spectrum*, is another airborne killer. Stomach analyses reveal that it regularly eats rodents and birds up to the size of doves. In captivity, it greedily consumes chickens and also its large, similarly carnivorous relative, the spear-nosed bat *Phyllostomus hastatus*.

Fish-eating bats

At least two species of bats feed mainly on fish. They both live in Central America; the fisherman bat (*Noctilio leporinus*) occurs as far south as Argentina, but the other species (*Pizonyx vivesi*) has a more restricted distribution. It also belongs to a different family and it is relatively poorly known.

Some insectivorous bats, such as the European Daubenton's bat (*Myotis daubentonii*), regularly fly low over water to catch newly emergent mayflies and it is easy to see how this habit might have evolved into fish catching. Fisherman bats find their prey by echolocation, but there are serious physical problems associated with their presumed habit of hearing echoes in flight from underwater targets. Various theories were advanced to explain how this was achieved and considerable controversy ensued. We now know that *Noctilio* flies low over still water using its echolocation system to detect ripples caused by fish near the surface. It then plunges its feet into the water in the hope of scooping out a fish. Usually it misses, but if a fish is caught, the bat has long hooked claws that hold it firmly. The prey is then carried to a favourite feeding roost to be eaten; about 20 fish may be consumed in a night.

Occasionally, the bat misjudges its aim and falls into the water. For most bats this would be a serious accident, but *Noctilio* can easily take off again from the surface and it has short greasy fur from which water is easily shaken off. However, such mishaps, and also the habit of dipping into the water, provide a chance for fish to get their own back on this particular bat. There are records of certain carnivorous fish catching *Noctilio* during its fishing flights!

Ospreys and Fish Eagles

The majority of the birds of prey catch their food on the ground (see page 138), but they are very adaptable creatures and it is not surprising that a number of them have evolved fish-catching habits. The major fishermen among the raptors are the osprey and eleven species of fish and sea eagles.

The magnificent osprey

The osprey (*Pandion haliaetus*) is a beautiful bird of prey with a wingspan of about 150cm (59in). It can be found almost all over the world, but it breeds mainly in the northern hemisphere – in North America, northern and eastern Europe, and northern Asia. In the southern hemisphere it breeds only in Australia. In America the osprey is known as the fish-hawk, although it does not belong to the same family as the true hawks.

Ospreys catch and eat a few small wading birds, but 99 percent of their food consists of fish – from both fresh and salt water. Hunting takes place only from the air, with the bird gliding or soaring in circles between 30 and 60m (197ft) above the water. When a fish is seen, the osprey may hover momentarily and then it drops almost vertically to the water. Its wings are partly folded and it leads with its head at first, but the feet are thrust forward at the last moment and the bird enters the water feet first with a considerable splash.

The preferred prey are fishes that swim very close to the surface, but the osprey can go further into the water than other raptors and can capture fishes as much as 70cm (28in) below the surface. In such instances the bird's wingtips are all that can be seen above the water, and even these may disappear on occasion.

The feet are very strong, with short, thick toes that can withstand the high-speed impact with the water, and the fleshy cere on the top of the beak can extend forward to close the nostrils and prevent water entering. The talons are long and strongly curved to grip the prey, while the outer toe can be swung backwards so that the grip is more symmetrical, and therefore stronger, with two toes on each side of the fish. Pads of sharp spicules under the toes provide further security by digging into the fish and preventing it slipping from the feet. Having secured its prey, the osprey rises rather untidily from the surface, shakes the water from its feathers, and flies off with the fish slung head first underneath it to reduce drag.

The prey is taken to a perch to be eaten, but the osprey is often robbed by fish eagles and sea eagles (see below). It may have to catch three or more fishes before it manages to keep one, and with an average of one fish caught for every four dives the bird may have to dive twelve times to get one meal. The whole fish is eaten except, perhaps, the head and the largest bones, and everything is digested apart from the scales, which are later cast up as pellets.

Scotland's famous ospreys – probably the most watched birds in the world – feed mainly on pike and trout in their Speyside breeding site. Other prey recorded in Britain include carp, bream, roach, perch, and small salmon. Goldfish have even been taken from ornamental ponds, and the ospreys frequently fly to catch grey mullet shoaling in coastal waters. But in their winter haunts, in the tropics, ospreys feed on hundreds of different kinds of fish.

The osprey is reputed to catch fishes weighing up to 1·8kg (4lb), but this must be rare because the bird itself averages only about 1·5kg (3·3lb). Prey taken in Scotland averages about 380gm (13·3oz), and the bulk of the fishes caught in Africa weigh less than 500gm (17·5oz). An adult osprey probably needs about 150gm (5·25oz) of food each day and, as a pair normally share their kills, a fish weighing 380gm would normally feed a pair quite easily. In practice, this often means two smaller fishes, and during the breeding season the male may have to catch five or six fishes each day. This does not seem to present much difficulty in northern waters, where piracy by the sea eagle is not particularly common, but it might explain why the osprey does not breed in Africa, where it is continually being robbed by the larger African fish eagle in search of an easy meal.

Spiny toes
Above The long, curved talons of the osprey take a firm grip on its slippery prey, helped by spines under the toes.

Snatching the prize
Left The osprey flies down to scoop a trout from the surface waters. Its heavily-built feet withstand the impact and close around the fish in a shower of spray. Quickly shaking the water from its feathers, the osprey flies off with its prize.

Sea eagles and fish eagles

The sea eagles and fish eagles form a rather distinct group, probably more closely related to the scavenging kites than to the other eagles. They have larger wings and shorter tails than typical eagles, and when soaring the wings and tail seem to merge into a single unit. All are more catholic in their feeding habits than the osprey and, although fish remain their preferred prey, many of them regularly eat carrion and catch various mammals.

The most famous member of the group is the bald eagle (*Haliaeetus leucocephalus*), which is the national emblem of the United States. It is not really bald and it gets its name for its snowy white head. Its fishing behaviour is quite variable; sometimes it will circle over the water and plunge down to snatch prey just like an osprey, but at other times it fishes from a perch. Dead and dying salmon provide it with abundant food after their spawning sessions and, although a rare bird, it can be seen in quite large numbers at such times in parts of Alaska. When fish are scarce, the bald eagle will kill rabbits and water birds and it will also scavenge at rubbish tips. Like its relatives, it also indulges in piracy against ospreys and other fish-eating birds.

Although the bald eagle sometimes plunges into the water like an osprey, the fish-eating eagles generally sweep their prey up with their talons at the end of a relatively slow glide. They concentrate on fishes that swim or bask near the surface and, like the osprey, they have sharp spicules on their toes to improve their grip. Northern sea eagles, such as the European sea eagle (*Haliaeetus albicilla*), have more powerful feet than their tropical relatives, associated with their tendency to take mammalian prey as well as fishes. Hares and baby seals are sometimes killed, and the European sea eagle has also been accused of killing sheep. It certainly does eat sheep, but probably only when it finds them already dead. Carrion of all kinds plays a considerable role in the diet of this eagle, and it also gets a certain amount of food by piracy – chasing ospreys on occasion until they drop their prey through exhaustion. The eagles may even snatch fishes from the ospreys' talons. Cormorants and other diving birds are quite easy prey; their reaction when they see the approaching eagle is to dive beneath the surface, but the eagle can watch them under the water and pounce as soon as they re-surface.

The African fish eagle (*Haliaeetus vocifer*) almost always hunts from a perch, and we know a great deal about it as a result of Leslie Brown's studies in Kenya. Spotting fish within about 50m

Tearing up fish prey with a hooked beak
Above This young African fish eagle is already an efficient hunter, capturing slippery prey with its taloned feet and tearing off strips of flesh with a cruel, hooked beak.

(164ft), the bird streaks down onto it, scoops it up, and returns to a perch in less than half a minute – sometimes in just 10 seconds. Catfish and lungfish, which regularly come to the surface for air, are among the eagle's commonest prey. When the water is rough and it is difficult to see fishes from a low angle on a perch, the eagle may soar above the water like an osprey. Its efficiency is much less than that of the osprey, however, and observations suggest that only one in ten pursuits are successful. Nevertheless, the fish eagle spends relatively little time fishing; up to 90 percent of the day is spent simply sitting on a perch, and a good deal of the rest may be devoted to piracy. Pelicans, herons, kingfishers, ospreys, and even other fish eagles are attacked and robbed, and yet the fish eagle virtually ignores carrion – a source of food that most other piratical birds are only too eager to accept.

The Diving, Plunging Seabirds

Seabirds belong to many different families and have many different habits but, with the exception of most of the gulls, they all rely on the intricate food chains of the water to provide them with their food. Even when they come ashore to breed they have to make regular trips to the sea to gather food in the form of fish, squid, and surface-dwelling crustaceans. Some of these seabirds, such as the penguins and the auks, habitually feed by diving from a swimming position on the water surface, and these birds are described on page 180. Here, we are dealing with those birds that use air-to-sea methods of hunting, whether they dive right into the water or simply snatch food from the surface as they fly along.

The graceful gannet

Gannets are goose-sized birds inhabiting cool seas. There are three species: *Sula bassana* of the North Atlantic, *S. capensis* from South African waters, and *S. serrator* from Australia and New Zealand. They are all very alike, and some ornithologists consider them to be merely varieties of one species. Like most seabirds, the gannets come ashore only for the breeding season, when they nest in vast cliff-top colonies. One cannot help admiring their graceful appearance as they soar out on their long, narrow wings, but their dives are even more spectacular.

The birds normally scan the water from a height of about 30m (98ft) and easily spot fishes or squid swimming near the surface. Having detected the prey, a gannet adjusts its position (see below) and goes into a near-vertical dive, with its wings folded back to give a beautifully streamlined shape like a swept-wing fighter plane. As the bird nears the water, it pulls its wings back even further, and when it enters the water the wings stream behind like a tail.

Entry speed may approach 160kph (100mph), and it is thought that the shock wave produced by the impact may stun the prey and give the gannet a better chance of catching it. The bird does not dive directly at the prey and spear it with the beak; it dives below the prey and, using both wings and feet for propulsion, swims up to grab its victim in the beak. If the prey is not stunned, the gannet may actually swim after it for a short distance, but the dive rarely lasts for more than about 10 seconds and the bird rarely goes deeper than about 15m (49ft). The prey may be swallowed as soon as it is caught under the water, or it may be carried to the surface. The gannet then flies up for another dive.

The main food of the Atlantic gannet, also known as the Solan goose, consists of herring, mackerel, and other shoaling fishes, together with large quantities of squid, but the bird will take any kind of fish that it can catch.

Good binocular vision (see page 68) enables the gannet to judge the depth at which its prey is swimming, and the bird adjusts its height so that it will dive to the right depth. The deeper the prey, the higher the bird flies before beginning its dive, and fishermen have often used the gannet as a guide to the depth at which to set their nets. When the shoals of prey are very close to the surface, the gannets may actually land on the water and fish cormorant-fashion by diving from there, but this behaviour is rare. Gannets are very buoyant birds because of their subcutaneous air sacs (see below) and they find it difficult to plunge beneath the surface without the impetus of a

The diving fisherman
Left The gannet will dive from 30m (98ft) into a shoal of mackerel. Folding its wings and extending its dagger-like beak, it plunges smoothly into the water after its prey. Swimming a short way under water, it grabs a fish and rises into the air.

Special adaptations for the soaring searcher
Above The gannet (*Sula bassana*) soars effortlessly above the sea on its long, slender wings as it searches for shoals of fish. Its blue-rimmed eyes are positioned to give excellent binocular vision as the bird looks down at the water. It has no sense of smell, for its nostrils are covered by bony plates to prevent the entry of water during a dive. The beak is stout and the front of the skull is also thickened to withstand the impact as the bird hits the water. Air trapped in its downy feathers as well as in special air-spaces beneath the skin cushions and spreads the impact forces. These air pockets also give buoyancy, however, and the gannet needs the impetus of a high dive to overcome it and get down to the fish.

steep dive from high above the water.

Gannets could not possibly dive into the water at speeds of up to 160kph without special modification to cushion the impact, even though they are so streamlined. There are, in fact, two major modifications, involving both the skeleton and the soft tissues. The skull, which takes the full force of the impact, is specially thickened to protect the brain, while the whole front end of the body is protected by a network of air pockets under the skin. These pockets are connected to the air sacs, which all birds possess as offshoots of their lungs, and they form very efficient shock absorbers for the head, neck, and breast. The birds' breathing arrangements have also been modified so that water is not forced into the lungs on entering the water. The external nostrils are covered by a sheet of bone, and the birds breathe through small gaps at the back of the beak.

The boobies
The boobies are the tropical equivalents of the gannets, and the six species are all placed in the same genus as the gannets. Their feeding behaviour is identical to that already described, although the prey species differ. Flying fish are among the commonest prey, although they are not taken while flying through the air. The boobies dive down into the water and come up to meet the flying fish as they land back in the sea. The Peruvian booby (*Sula variegata*), which is one of the famous guano birds, feeds largely on anchovies. Boobies carrying food back to their nests are very often attacked by frigate birds (see below) and made to disgorge the food. The gannets of cooler waters are similarly attacked by the piratical skuas (see page 151).

The brown pelican
Seven of the eight species of pelicans are primarily birds of freshwater and they generally feed by swimming on the surface or by wading in shallow water (see page 184). The brown pelican (*Pelecanus occidentalis*) is a seabird, however, and it has adopted a very different method of hunting. It dives for fish in a manner almost identical to that used by the gannets and boobies, and it has the same anatomical modifications to protect it from injury when it hits the water. It is not such a neat diver as the gannet, coming down with wings half open and with its head pulled back into its shoulders, and it makes a considerable splash on entry. The noise is said to be audible nearly a kilometre away on a calm day, and the shock wave is believed to stun fishes as much as 2m (6·5ft) below the surface. Several fishes may be scooped up in the pouch before the pelican surfaces again, and then the bird has to tilt its head to pour the water out before it can swallow the fishes.

Thousands of brown pelicans live on the guano islands just off the coast of Peru. Many boobies live here as well, but competition between the species is largely avoided because the pelicans do not dive as deeply as the boobies and do not go so far out to sea. Brown pelicans around the southeastern coasts of the United States were once accused of taking commercially valuable food fishes, but a careful study revealed that they were feeding mainly on an oily fish known as the menhaden (*Brevoortia tyrannus*). This has no value as human food, although it is increasingly caught by today's fishing fleets to provide fish meal and oil.

The piratical frigate birds
The frigate birds (*Fregata* spp) manage to be seabirds without ever being water birds in the accepted sense. They never enter the water deliberately, and if they do land on the water they are almost certainly doomed to die, for their legs are so short and their wings are so long that they cannot get themselves into the air again. Such accidents are probably very rare, however, for the frigate birds are among the world's most accomplished aeronauts. Audubon considered their powers of flight to be 'superior to that of perhaps any other bird.' They weigh less than 2kg (4·4lb), and yet their wings span more than 2m (6·5ft), giving them a perfect build for high-speed manoeuvres and effortless soaring.

The three species of frigate birds breed on tropical coasts and islands and, because they cannot rest on the sea, they rarely fly more than about 100km (62 miles) from land. Even on land, they have to rest in trees or on high rocks so that they can get airborne again. Much of their food consists of flying fishes, which they snatch with amazing precision from just above the waves. The fishes 'fly' for about 10 seconds at a time, reaching speeds of about 60kph (37mph), and the frigate birds are able to detect and overhaul them in this time. The birds often position themselves over shoals of tuna and other predatory fishes. Flying fishes take to the air in an attempt to avoid these predators, and the frigate birds swoop down and have a real feast.

The brown pelican
Above right Two brown pelicans fighting over a fish. Such fights are probably uncommon, for fish is usually plentiful in the birds' natural habitat. The bird on the right is gliding through a shoal of flying fishes.

Several other small fishes are snatched from the surface waters, especially in rough weather, when the frigate birds actually swoop along the troughs between the waves. Hatchling turtles are snatched from the sand with incredible speed and accuracy, the birds' wings rarely even touching the ground as they swoop down.

Offal from ships is avidly snatched up, often before it hits the water, but it is their piratical attacks on the boobies that have made the frigate birds famous and given them their alternative name of man-o'-war birds. Boobies carrying food back to their nests are harassed by the men-o'-war, which often work in groups and peck at the boobies from all sides. The torment continues until the boobies disgorge their food in an attempt to lighten their weight and escape; the frigate birds immediately break away and streak after the falling food, grabbing it easily before it reaches the water.

Frigate birds seem to know straight away whether a booby is worth attacking – perhaps because a laden bird flies more slowly or because it calls out at the first sign of attack and the attackers can distinguish between the call of an empty bird and one full of fish. Piracy clearly pays best during the breeding season of the boobies, when large numbers of them occur in one area and all are busy carrying food to their nests. Attacks are made at other times of the year, but they are less frequent and the frigate birds resort to hunting at first hand.

Frigate birds rob boobies of sticks and other nesting materials on occasion, although there seems to be even less reason for this than for plundering food. The frigate birds nest in the trees and their agility in the air makes it very easy for them to pluck twigs as they fly by.

The food-catching abilities of the frigate birds have to be learned by the youngsters and this often takes a long time. The parent birds may continue to feed the young for six months after they have left the nest, during which time the young gradually improve their techniques by swooping after falling twigs and leaves. Many of them fail to master the techniques sufficiently, however, and large numbers die as soon as they are left to fend for themselves.

Skimming the surface

The three species of skimmers or scissor-bills (*Rhynchops* spp) form a small family of seabirds related to the gulls and terns (see below). The birds actually look like heavily-built terns except for the abnormally large and asymmetrical beak. The latter is the basis of their unique method of fishing while skimming low over the

Swooping pirates
Right Frigate birds soar above the seas on their long wings and watch for boobies returning with catches. When a victim is selected, the fast-flying frigates fold their angular wings and swoop down to mob the heavily-laden booby. In its effort to escape, the slower-flying booby disgorges its catch, and the agile frigates snatch the food in mid-air. Such piratical attacks are most common during the boobies' breeding season.

From their tropical oceanic islands, the piratical frigate birds also fly out to sea to catch their own prey. Sweeping low over the waves, they snatch surface-swimming fish, squid, and crustaceans without wetting their glossy black feathers. Only their long, hooked beaks dip into the water to close around the slippery prey.

surface. The lower half of the beak is much longer than the upper half and it plunges through the water like a scoop. The beak is held wide open while fishing, and the gape is such that the upper mandible remains well above the water. Drag is reduced to a minimum by the narrowness of the beak, and the relatively large wings generate the extra lift and thrust necessary to keep the bird moving forward.

As soon as the lower mandible touches something edible – a fish or squid, or perhaps a crustacean – the head is flicked down so that the beak is in a vertical position and the beak is simultaneously snapped shut. The knife-like edges of the mandibles ensure that the prey is firmly held, and then the powerful neck muscles raise the head so that the prey can be swallowed. Flight is not interrupted, and the beak is quickly returned to the flying position. The birds fish only in calm waters.

Skimmers do a lot of their feeding at dusk and even at night, and their eyes are remarkably cat-like – quite unlike anything found in other birds. The pupil is large and round at night, giving the bird good vision when it is hunting, but

The glistening lure of the night-flying skimmer
Below A skimmer shears through the water with the elongated lower half of its beak. On hitting a fish or crustacean, the beak snaps shut. The shining wake lures more fishes, and the bird retraces its path.

it contracts to a narrow, vertical slit during the daytime, thus preventing discomfort to the birds as they roost on brightly-lit sand banks.

The black skimmer (*Rhynchops nigra*), which lives around the coasts of North and South America, actually attracts a good deal of food with its activity. Disturbance of the surface waters stimulates the phosphorescent algae, and fishes and other creatures come up to investigate. The skimmer then turns back along its track and reaps its harvest. The black skimmer feeds only in the sea and in estuaries, but the African skimmer (*R. flavirostris*) and the Indian skimmer (*R. albicollis*) feed on inland waters as well. The Indian skimmer, which extends from India to Vietnam, rarely ventures out to sea.

When the skimmers hatch from the eggs the two halves of the beak are of equal length, and the chicks are able to pick up small food items, such as insects and crustaceans, from the ground. The lower mandible begins to grow when the chicks start to get their flight feathers, and the birds are able to skim food soon after they learn to fly, but the scoop is not fully developed until the skimmers reach maturity.

The terns

The 39 species of terns belong to the same family as the gulls, but they are much daintier birds, with pointed beaks, narrow wings, and forked tails. They have a more fluttering flight than the

The hovering, plunging fisherman

Below The graceful Arctic tern hovers over the sea before plunging onto its prey. Sharp eyesight locates the target, even in rough water.

gulls. The majority of the species inhabit the warmer seas, but terns are in no way confined to warm waters — the Arctic tern, for example, breeds way north of the Arctic Circle. Most of the species nest on the seashore and often fly far out to sea to feed, but a few live on inland seas and marshes. Five tropical species, known as noddies, nest on cliffs and in tree tops.

The terns do not swim well and do not often land on the water. They hunt by quartering the surface from a few metres above the waves, with the head down and the beak almost vertical. When food is seen, the bird may hover above it for a second or two to fix its position and then, closing its wings, it may drop right down on the prey. This is known as the plunge-dive, but the bird does not usually go right under the water. It grabs the prey in its beak and rises immediately, shaking water droplets from its feathers as it goes.

Not all terns go in for the plunge-dive, however; some of them, including the noddies, daintily pluck food from the surface while hovering. The food is generally swallowed at once, unless it is being taken back to feed a mate or young, in which case it is carried in the beak and the terns may have to face attack by skuas (see page 151). A good deal of courtship ritual involves carrying fishes in the beak.

The main food of terns consists of fish, squid, and crustaceans — all generally under about 10cm (4in) long. Flying fishes are an important item in tropical waters, terns often catching them by following shoals of tuna in the same way as frigate birds. Inland terns catch

large numbers of insects on the wing, while Forster's tern (*Sterna forsteri*) has adopted scavenging habits during its spring migration. It keeps pace with the thaw as it moves north and feeds on the numerous small animal corpses released by the melting ice.

Sea gulls galore

The most familiar seabirds are undoubtedly the gulls that haunt coastlines and follow ships out to sea. As a group, however, these birds are much less dependent on the sea than the species described above. A few species feed on fishes and other animals far out at sea — generally using the plunge-dive technique of the terns or feeding from the surface like ducks — but the majority are scavengers on and around the seashore. Some, such as the herring gull and the black-headed gull, feed far inland, especially during the winter. They find rich pickings on farmland and rubbish dumps, and there are now large populations of gulls in many areas of the world that never go to sea.

The gulls soar effortlessly in their search for food, often flying in large flocks and thus increasing the search area. When one bird sees something, it drops down and the others follow. Small fishes may be snatched from the water in the manner of frigate birds or terns, but the gulls more often land on the water and feed — especially when taking offal thrown from ships. On the shore, they examine the strand line for dead animals, and often carry shellfish into the air and then drop them in an attempt to smash the shells. This works well on a rocky shore, but the birds fail to realize that the shells will not break on a sandy beach and they make many futile attempts before giving up. Worms and other burrowing creatures may be obtained by treading the sand at low tide and forcing them to the surface.

Some of the gulls become pirates and predators of other seabirds during the breeding season. Herrings gulls, for example, bully puffins into dropping their catch, and even dive down to knock the puffins over on their breeding grounds. Before the puffins can recover, the gulls have taken their food. The great black-backed gull (*Larus marinus*) regularly drags puffin chicks from their burrows, and even catches and eats adults. It also pirates food from other gulls and, like many other gulls, frequently indulges in cannibalism by taking and eating the chicks of its own species — even its own offspring. Outside the breeding season the gulls have to rely more on fishing and scavenging, but there is still the possibility of piracy against other species of birds.

Kingfishers and Dippers

The most famous of the freshwater diving birds are undoubtedly the kingfishers, typified by the common kingfisher (*Alcedo atthis*) of Europe and much of Asia. There are about 80 species altogether, although they are not all fishermen, and the majority are brilliantly coloured. All have relatively large heads and beaks, well suited to catching fishes and other prey in the water.

The common kingfisher

The common kingfisher likes rather shallow streams and pools, especially those with gravelly bottoms. It is not uncommon around watercress beds. It will fish in deeper waters, but does not dive deeply and always waits for fishes to come within a few centimetres of the surface. Even 30cm (12in) is too deep for the kingfisher under normal circumstances. Although clear waters are preferred, the bird is not averse to fishing in muddy canals, and here it has to wait until the fishes almost break the surface before it can see them.

The kingfisher hunts almost entirely from a perch overlooking the water, and usually overhanging it as well. The height of the perch is very variable. Ron and Rose Eastman, who made the amazing film *The Private Life of the Kingfisher*, found that the bird fished the watercress beds quite happily from the concrete walls 25cm (10in) above the water, but that it chose perches 2m (6·5ft) or more above the water when fishing a fast-moving stream. This initially suggested that the kingfisher works from different heights over slow and fast streams, but height is actually less important than the rigidity of the perch and the birds generally choose the firmest supports. Kingfishers strung out along the muddy klongs or canals of Thailand use perches anything from 60cm to 4m (2–13ft) above the water, but always seem to select the thickest branches available.

An eye for a fish

A kingfisher on the lookout for a meal sits on its perch with its head slightly to one side and scans the water with one eye. From time to time it cocks its head the other way and uses the opposite eye. Any fisherman, or anyone who has sat in a small boat for any length of time, will know that one of the worst problems is glare from the water surface, but the patient kingfisher overcomes this problem by having a very high proportion of red oil droplets in its retina in addition to the yellow droplets found in the retina of all diurnal birds. The red oil droplets act as an efficient filter that counteracts glare and the kingfisher can thus see fishes just under the surface much more clearly than we can.

When it spots a fish, the kingfisher tenses itself and pulls its feathers in tightly to increase streamlining and water repellency. In a flash, it leaves its perch and plunges head-first into the water. The wings close for the moment of impact, but open again under water and the bird swims straight at the fish with its beak open. The prey is thus caught in the beak, and not speared by it. The bird flaps to the surface and flies back to its perch, the whole process being over in about a second when fishing from a low perch. Only about one third of a second is spent under the water catching the fish.

Unlike the dipper (see below), the kingfisher cannot change the shape of the lens in its eye to improve vision under the water, but it has an egg-shaped lens and two specially sensitive spots, or foveas, on the retina instead of the usual one. The light rays are concentrated onto the central fovea in air, but onto the lateral one in water, thus assuring the bird of good vision in both environments.

An adult kingfisher averages one catch for every three dives, although not every dive may be a 'serious' one in definite pursuit of prey. The captured prey is generally bashed against the perch to subdue or even kill it – hence the need for a firm perch – and then

End of the underwater flight
Above A common kingfisher emerges from the water with a sizeable fish held firmly in its beak. The bird actually 'flies' under the water by flapping its wings and it powers itself to the surface sufficiently rapidly to break through and rise into the air.

swallowed head first, often after a neat piece of juggling. The main prey in Britain are minnows, sticklebacks, and the bullhead, or miller's thumb, (*Cottus gobio*). Minnows are swallowed without any difficulty, even if still alive, but the other two fishes are rather spiny and must be killed before they are swallowed. Only when they are completely dead do the spines lay flat, and even then the

High-speed dive
Right Spotting a fish
from its perch, a
kingfisher pulls its
feathers in tightly and
plunges head-first into
the water. Wings and
beak open under the
water, and if the bird
is lucky it catches the
fish. It flaps quickly
to the surface and
flies up to its perch
with its catch. The
whole process may take
no more than one
second, and only one
third of this time is
spent actually under
the water. Returning
to its perch, the bird
has to subdue its prey,
usually by bashing it
against the perch, and
it then swallows it
head-first. The prey
is not torn up before
it is swallowed.

Portrait of the victor
Left A common
kingfisher stands
triumphantly on its
perch, but it has a
bit of juggling to do
to get its large catch
turned round so that
it can be swallowed
head-first.

fishes must be swallowed head-first if
the spines are not to stick in the kingfisher's throat.

Even nestling kingfishers swallow
whole fishes, forming an orderly queue
in the nest burrow so that each one is
fed in turn as the parent birds fly in
with their catches. Bones are regurgitated, and the adult birds use a good
number of them to line their nest.

Fishes are the dominant prey of the
common kingfisher, but a few other
types of food are taken on occasion.
Caddis larvae and other insects are sometimes caught in the water, and when the
kingfisher goes down to the coast – as it
often does in hard winters – it catches
a good number of shrimps and molluscs.
Occasionally the bird will dart after
dragonflies and other flying insects in
the manner of a bee-eater – a sort of
tit-for-tat, for some bee-eaters are known
to plunge into water for fishes. At least
one kingfisher had been known to steal
food from dippers surfacing with beaks
full of tasty morsels.

Kingfishers over the land
As already stated, not all kingfishers are
actually fish-eaters. Most of the tropical
species live in the forests and, although
they still hunt from perches, they swoop
down to snatch lizards and other small

A busy life for the parent dipper
Above As well as
collecting food for
itself, the adult
dipper has to satisfy
the ever-open beaks of
its young during the
breeding season. It may
bring hundreds of items
to the nest each day.
This adult is about to
give its offspring a
young water beetle.

animals from the forest floor. Some take
insects in mid-air. The most unusual,
however, is the shoe-billed kingfisher
(*Clytoceyx rex*), which uses its flattened
beak to dig for earthworms in New
Guinea. The most famous of the non-
fishing kingfishers is the Australian
kookaburra. It eats almost any animal
that it can catch in its big beak, and
has quite a reputation as a snake-killer.

The insatiable dipper
The dipper gets its name for its habit of
standing on a rock and bobbing up and
down, but it could equally well have been
named for the readiness with which it
takes a dip in the water. The common
dipper (*Cinclus cinclus*) of Europe and
Central Asia has been known to take
1,600 dips in a day, spending as much
as two hours under the water altogether
and covering a total distance of nearly
2km (1·25 miles). Each dip is, of course,
a feeding run, for the dipper collects
almost all of its food under water.

There are four species of dipper altogether: the common dipper, the North
American dipper, the white-capped dipper from South America, and the brown
dipper from south and east Asia. All are
very similar in appearance and habits.
They are the only truly aquatic perching
birds, and they have very dense, oily
plumage. They feed in swift-flowing,
rocky streams, preferring those in which
the water is no more than about 60cm
(24in) deep. The usual feeding method
is to walk down a boulder and into the
water and then climb out onto a neighbouring boulder. It used to be thought
that the bird actually walked along the
stream bed, but once under the water it
uses its wings to 'fly.' It may use its feet
as well when close to the bottom, but it
often feeds in mid-water and then relies
entirely on its wings. The bird sometimes dives from its perch, and it can
also find food by wading in the shallows.

The dipper's eye has an extremely
soft lens and some very strong muscles,
which together provide an amazing
degree of accommodation for underwater
vision. On entering the water, the bird
immediately loses the focusing power of
the cornea of the eye, but the lens is
made to bulge enormously to compensate. The bird can thus see as well under
water as it can in the air. The same
system is found in the cormorants (page
180) and probably in some other diving
birds, but not in the penguins. These
birds see very well under the water, but
have poor vision on land.

Dippers pick up a wide variety of
aquatic creatures while under the water.
Fishes, molluscs, worms, and various
insect nymphs and larvae are all eaten,
but perhaps the most important foods
are small crustaceans, such as the freshwater shrimp (*Gammarus pulex*). These
make up more than 80 percent of the
dipper's diet in some places, and several
can be seen in the beak when the bird
climbs out of the water. The dipper has
a great appetite and may eat its own
weight of food each day – perhaps
75gm (2·6oz) of assorted aquatic creatures; no wonder the bird makes so
many dives each day.

Hunters in the Water

Vibrations and electric fields augment the more usual
food-finding senses of sight, scent and hearing in many aquatic predators. Electric shocks
join powerful teeth and suckers as the major weapons.

The aquatic environment provides the opportunity for one method of nutrition denied to land-dwellers. This is filter-feeding, in which an animal takes in large quantities of water and then pumps it out through some kind of filter or sieve that holds back food particles for the animal to eat. The particles are usually very small in relation to the animal that eats them. Most filter-feeders are scavengers, exemplified by the bivalve molluscs that live on the seabed and sieve through the debris that rains down on them. These have nothing to do with hunting, but it is not so easy to dismiss all the filter-feeders. What about the great whalebone whales (page 204) that filter tons of planktonic animals from the water? These are clearly predators and, although they do not literally chase their prey, they could be classed as hunters on the basis that they move about and take living prey from their environment.

The nearest approach to filter-feeding that we find on land – and it is not a very close approach – is the way in which nightjars gather gnats as they fly through swarms of these insects at dusk (see page 154).

Detecting prey under water

Turning now to undisputed aquatic hunters, it is clear that all the major senses – sight, hearing, and smell – are used to detect prey, as well as a number of rather specialized ones. The sense of taste, which is confined to the mouth in land-living vertebrates, is much more diffuse in fishes, with taste buds often scattered all over the body. Barbels, catfishes, and some others possess fleshy 'whiskers' with heavy concentrations of taste buds. These help to detect food as they are dragged over the river bed.

Low-frequency vibrations play a major role in the aquatic environment, giving both predators and prey information about other creatures in the vicinity. Fishes and some other aquatic vertebrates pick up the vibrations with a special piece of equipment known as the lateral line system (see page 71). Seals and otters use their stiff whiskers, or vibrissae. These stout hairs have elaborate networks of nerves around their bases and they can detect the slightest disturbance of the water. They can even pinpoint the direction from which the disturbance comes, and thus lead the predators towards passing prey. The tiny arrow worms (*Sagitta* sp) that live in the plankton also find their food through vibrations; sensitive bristles around the jaws detect the tiny pressure waves set up by a passing crustacean and the arrow worm darts forward to grab it.

The most unusual sesnses displayed by aquatic hunters are the electrical senses of various fishes. It has long been known that some fishes can give out powerful electric shocks (see page 190), but as more species are being studied it is becoming clear that a great many use electricity for navigation and for prey detection. They generate electric fields around themselves and detect any interference with these fields such as would be caused by approaching prey.

The chase

Having detected potential prey, the aquatic hunter must give chase. Some bottom-living animals can indulge in stalking techniques like those we have seen on land, but speed and manoeuvrability are the main features needed for chasing active prey under water. Camouflage is also important for both predator and prey. Disruptive coloration, with bold stripes breaking up the outline, is found in many freshwater fishes and also among those inhabiting coral reefs. Pelagic, or free-swimming, species tend to exhibit strong countershading, with the upper surface much darker than the belly. This is true of penguins and many marine mammals as well and it confers advantages from every direction. Viewed from the side, a counter-shaded fish becomes 'flat' and merges with the surroundings because light shining from above lightens the back, while the pale colour of the belly counters the shadows below. Seen from above, the dark back of a fish merges with the dark waters below, and when seen from below the silvery belly blends beautifully with the light shining down from the surface. It is not surprising that few pelagic predators rely entirely on sight to find prey.

Quite a number of aquatic hunters search by actively flushing out their prey and catch it before it has a chance to escape. Otters, for example, often stir up the sediment on the river bed to disturb crayfishes and other prey. Phalaropes and other birds do the same, although they are generally searching only for slow-moving prey such as worms and molluscs that live in the mud.

Aquatic weapons

Apart from the electric shocks handed out by some fishes, almost the only weapons available to the aquatic vertebrates are their teeth and jaws, but a look at the teeth of a barracuda (page 188) or a crocodile (page 198) shows that such weapons can be more than adequate for their purpose. The invertebrate hunters have a greater variety of weapons, but they still rely largely on their jaws and associated equipment. The dragonfly nymph, for example, uses a much modified lower lip to impale prey (see page 178), while many carnivorous molluscs (see page 174) use the rasp-like radula, or tongue, to drill into their victims. The cephalopods (see page 196) capture prey with the aid of their sucker-covered tentacles, but the final touch is added by a poisonous bite from the horny beak. Crabs and lobsters capture prey with their great pincers, but the starfish uses tiny suckers to overcome its victims. Each sucker is very small, but a starfish has a great many of them and their combined pull enables it to open bivalves with ease (see page 176).

Perhaps the most unusual weapon among the aquatic hunters is the proboscis of the nemertines, or ribbon worms (phylum Nemertea). There are about 600 known species and most of them live in the sea. They range from a few centimetres to about 50m (164ft) in length, but they are all very thin and often flat. The proboscis is like a miniature harpoon which, in some species, is as long as the body. It is stored in a sheath and fired out to capture prey in much the same way as the thread cells of the sea anemones and their relatives (see pages 38 to 43).

Some species have a poisoned spine at the end of the proboscis, and this is driven into the prey to subdue or even kill it. Other species lack the spine and the proboscis merely wraps around the prey and holds it while the ribbon worm sucks it, whole or piecemeal, into its mouth. Ribbon worms feed mainly on other worms and molluscs, but they sometimes catch fishes. Unlike the coelenterate thread cell, the ribbon worm proboscis can be retracted and used again whenever necessary. It is fired out whenever the worm bumps into something that stimulates its sense of taste or smell in the right way.

The mask of death

Right A weed-dwelling dragonfly nymph (Aeshna cyanea) has fired out its mask (see page 178) and impaled a froglet on the claws. The mask has now been retracted so that the jaws can get to work on the prey. Large eyes detect prey up to 20cm (8in) away.

Drilling for Food

All the slugs and snails belong to the large group of molluscs known as the gastropods, and they all feed primarily by means of a piece of apparatus known as the radula. This is rather like a ribbon with many transverse rows of horny teeth and it is supported on a strip of tough, cartilage-like material. The whole thing, together with its associated muscles, is called the buccal mass and it can be pushed forward to the opening of the mouth, where the radula teeth are used to rasp away the animal's food. The teeth at the front wear out quite quickly, but they are continually being replaced as new ones are formed at the back and the whole radula moves forward on its support. The radula teeth are very variable and can be used for simple rasping, for drilling, or even for harpooning moving prey. Many of the carnivorous sea snails carry the mouth and radula at the end of an extensible proboscis, which may be as much as five times the length of the body when fully extended.

Planktonic predators

Most snails are rather slow-moving creatures, restricted to a life of crawing over the ground or vegetation, or perhaps burrowing into soft ground. The sea snails are no exception to this general rule, but a few species have managed to break away from life on the seabed and now live freely in the surface waters.

One of the best known of these is the violet snail (*Ianthina janthina*). It floats upside down attached to a raft of bubbles which it forms by trapping air in mucus from its foot, and if it ever becomes de-

tached from its raft it is doomed to sink to the bottom and die. *Ianthina* lives in most of the oceans and, although it has no eyes and no means of controlling its drift across the waves, it is an effective predator of *Velella* (see page 42). Both predator and prey drift with the wind and thus frequently bump into each other. *Ianthina* then pushes its short proboscis against its prey and takes a firm hold with its strongly curved radula teeth. The purple dye exuded by *Ianthina* may poison the prey, and the snail then rasps it to pieces with its radula.

The heteropods, typified by *Carinaria mediterranea*, are free-swimming gastropods in which the foot is laterally compressed to form a fin. The shell is reduced, or even absent, and the animals swim in an upside-down position by 'rowing'

with the single fin. The eyes are large and used to detect other small animals that are used for food. The prey is caught by the massive, trunk-like proboscis and quickly clawed to pieces by the radula.

Drilling for food

Quite a number of carnivorous molluscs, such as the cowries, simply browse on sponges, sea squirt colonies, and other sedentary animals, in much the same way as the limpets browse on algae on rocks. Many other species have to work harder for their food. The necklace shells of the genus *Natica* are among many that feed on bivalve molluscs, and they have to drill through the shells before they can eat. *Natica* uses its great wedge-shaped foot to plough through the sand as it searches for the bivalves. When it finds a suitable victim, it envelops it with the foot and brings the proboscis into play. If the bivalve is not completely closed, the radula gets to work to rasp the flesh immediately, but the predator's approach generally causes the bivalve to close. In this case, a large gland on the underside of the proboscis pours a softening fluid onto the bivalve shell in one small area. After a while, the radula can drill away the softened material. Unless the bivalve is very young and its shell is very thin, several applications of softening fluid and several sessions of drilling are necessary to break right through, and then the proboscis is pushed right into the bivalve and the radula demolishes every scrap of flesh.

The molluscs described so far belong to the group known as mesogastropods, but the most highly developed carnivores belong to the group known as the neogastropods. The radula of these carni-

Ianthina – the predator in a raft of bubbles
Below The violet snail, or bubble-raft snail, floats at the surface of the sea in a raft of bubbles. It feeds on various coelenterates, such as *Porpita* seen here, and *Velella*.

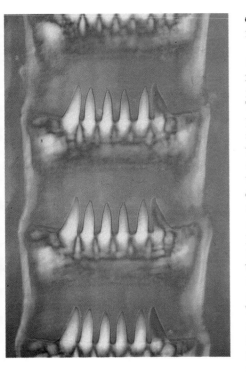

Colourful whelks
Above left A group of dog-whelks feeding on mussels. Part of the rock has been cleared, but the mussels' anchor threads remain.

A rasping tongue
Above This very highly magnified picture shows just a few of the rows of rasping, horny teeth to be found on a whelk's radula.

vores has no more than three teeth in each row, but each tooth may have many points and the radula remains a very efficient weapon. It is carried at the end of an extensible proboscis that can be shot out to seize and also poison active prey considerably larger than the snails themselves. The proboscis is fired out by blood pressure and retracted by muscles. Prey is scented by means of an organ known as the osphradium, which is situated close to the gills. It 'tastes' the water drawn in through the siphon for breathing purposes and if it detects the scent

of prey the animal turns in the direction from which the water is being drawn.

The common whelk (*Buccinum undatum*) is probably the best known of the neogastropods, but it is not typical of the group in that it feeds largely on carrion. The dog-whelk (*Nucella lapillus*) is a real predator, specializing in barnacles, mussels, and limpets. These prey animals all thrive on wave-battered rocks, and the dog-whelk has evolved a very thick shell that enables it to withstand the rough conditions. Barnacle shells are prised open by the dog-whelk's proboscis, probably aided by a poisonous secretion called purpurin that paralyzes and relaxes the prey's muscles. Limpets and mussels have to be drilled by the proboscis, using the same combination of chemical and mechanical methods that we saw in *Natica*. The dog-whelk may take two days to drill through a mature limpet or mussel shell, during which time it moves its position many times. The radula thus works in many different directions and eventually produces a very neat, circular hole. A good deal of research has been done on the softening fluid produced by the various shell-borers, but its nature is still not really understood. It does not seem to be an acid, which was the original suggestion.

Other familiar shell-boring sea snails include the oyster drill (*Urosalpinx cinerea*), which is a serious pest in oyster beds, and the various *Murex* species whose beautiful, spiny shells are much sought after by shell-collectors. Some of these shells have a special tooth on the rim and they use it to prise open various bivalve shells.

The deadly cones
The most highly specialized of the carnivorous sea snails are the cones (family

Conidae). These are tropical creatures, with shells reaching 15cm (6in) or so in length, and they feed on worms, other sea snails, or fishes. The animals themselves do not move very quickly, but they can extend the proboscis quite violently when they come into contact with suitable prey. The scent of prey picked up in the water causes the cones to search around, and when the siphon has fixed the position of the prey the proboscis is fired. It may equal the length of the body, and it is a very bulbous affair with a poisonous harpoon at the tip. The harpoon is actually a much-modified radula tooth, grooved and barbed and up to 1cm (0·4in) long.

A venom sac is also connected to the proboscis, and when the latter is fired poison runs through to the tooth and into the victim. The prey is quickly paralyzed, and then it is engulfed by the proboscis and digested there. This may take several hours when the prey is as large or larger than the cone itself, and the mollusc cannot withdraw into its shell until digestion is complete.

The harpoon is discarded after use, and the radula moves another one forward to take its place at the tip of the proboscis, ready for the next victim. The detailed form of the harpoon varies from species to species and differs according to the preferred prey. The venom also varies slightly according to the type of prey taken by the cones, but it is chemically related to curare and it has a rapid paralyzing action on the muscles. When used against other gastropods it makes them release their hold on their shells, and the cone can then engulf the soft parts only. Several of the fish-eating cones have particularly dangerous and potent venoms and have been known to kill people on several occasions.

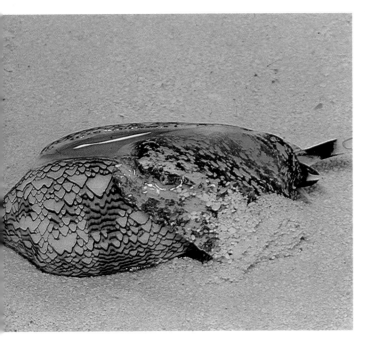

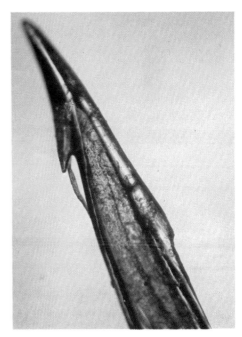

The textile cone
Left Beautiful, but deadly, the textile cone (*Conus textile*) is here seen in its attack position, with the proboscis on the right of the picture ready to be fired. This species, which comes from the Indo-Pacific region, has a most attractive shell, which reaches a length of about 10cm (4in).

The cone's harpoon
Right A much magnified radula tooth from the textile cone. It is clearly barbed, and when it is fired into the prey it becomes firmly embedded in the tissues. Poison is pumped through it and into the prey. Each tooth is used only once and a new one quickly replaces it after use.

The Starfishes – Slow but Sure

The starfishes belong to a very ancient group of animals known as the echinoderms. Other members of the group include the sea lilies and feather stars, the brittle stars, the sea cucumbers, and the sea urchins. All of them are confined to the sea, some of them living close to the shore and others in deep water.

The sea lilies and feather stars use their delicate, spreading arms to catch particles of debris falling onto them. Brittle stars and sea cucumbers generally feed on debris that they scoop from the seabed. Some burrowing sea urchins, such as the sea potato (*Echinocardium cordatum*), also feed on detrital particles in and on the sand, but most sea urchins are vegetarians. They use a complex piece of equipment known as Aristotle's lantern to scrape algae from the rocks. The lantern consists of 40 hard 'bones' and a mass of muscles, all designed to operate five projecting teeth in the manner of a grab to pluck or drag the plants from their homes. Encrusting animals, such as bryozoans, may be scraped up as well, but the urchins cannot be regarded as carnivorous. Only among the 1600 or so species of starfishes do we find truly predatory habits.

Water power

Starfishes, with their five or more arms radiating from a rather indistinct central disc, are familiar animals of the seashore, but few people know how they move about. A look at the underside of a living specimen will reveal the answer. Covering the lower surface of each arm there are masses of delicate, waving tubes known as tube-feet. There may be several thousand of them to each starfish, all interconnected to form the water vascular system. This system is found in all echinoderms, but not in any other animals. Each tube-foot is filled with water, which is normally under just sufficient pressure to keep the tube inflated. Pressure can be built up in the tube-foot, however, by the contraction of a small bulb, or ampulla, embedded in the arm at the top of each tube. The increased pressure extends the tube-foot in any desired direction.

In most starfishes each tube-foot ends in a little sucker, and when this sucker is pressed against something a tiny muscle inside it contracts to arch up the sole and form a suction cup. The result is a very firm grip. The starfish has only a very simple nervous system, but it can

The queen scallop escapes in a flap
Above Most molluscs are easy prey for the starfish, but many scallops can swim away from the danger by flapping the shells.

co-ordinate the movement of the tube-feet and, by extending them all in one direction to take a grip and then contracting the tubes all at once, the starfish can drag itself along. Movement is rather slow, but the starfish's prey is slower still, or even completely sedentary, and the starfish does not need to be speedy to be an efficient predator.

The tube-feet also absorb much of the starfish's oxygen requirement from the surrounding water and, as we shall see, they play a major part in both finding and absorbing the animal's food.

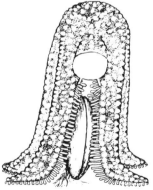
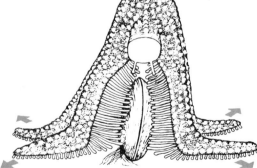

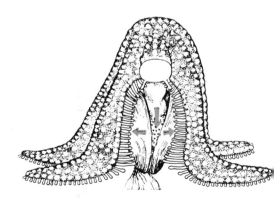

Opening a mussel by suction power
Above The starfish crawls over a mussel and drags it upright so that the hinge lies along the ground. One arm of the starfish has been cut away in the diagram so that the rest of the process can be illustrated. The suckered tube-feet are firmly applied to each valve, and the tips of the starfish's arms spread out on the seabed (centre picture), causing the tube feet attached to the mollusc to elongate. With the tips of the arms firmly anchored, the tube-feet then contract and gradually pull the valves apart. The starfish can then begin to digest its prey.

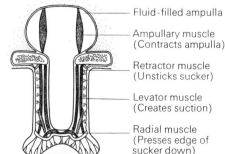

— Fluid-filled ampulla
— Ampullary muscle (Contracts ampulla)
— Retractor muscle (Unsticks sucker)
— Levator muscle (Creates suction)
— Radial muscle (Presses edge of sucker down)

Water-powered feet
Left The detailed structure of a tube-foot. Contraction of the ampulla extends the tube-foot, while contraction of the levator muscles raises the centre of the disc and creates the suction. Other muscles move the whole tube-foot. *Right* The tube-feet of a walking starfish.

The starfish and the mollusc

The typical starfish, such as *Asterias rubens*, that we find on the seashore feeds on a variety of other animals, both living and dead, but its main foods are various kinds of molluscs. The starfish has no proper eyes, although each arm has a light sensitive patch at its tip, and the animal has to rely on its chemical senses to find its food. Many of the tube-feet are sensitive to chemical stimulation, but the most important in this respect are those at the tip of the arms and those around the mouth, which is in the centre of the animal's lower surface. The starfish thus literally has to crawl onto its prey before it recognizes it, but such is the density of molluscs on the seabed that it is most unlikely that the predator has to go hungry for very long.

Small molluscs are swallowed whole, being shovelled into the mouth by the surrounding tube-feet. The soft parts are digested, and the shells are later ejected. Prey that is too large to be taken into the starfish body is dealt with in a most unusual way: the starfish everts its stomach and wraps its tightly round the prey. Digestive juices are then secreted and the soft parts of the prey are dissolved. Very little of the fluid is lost because the stomach is wrapped so tightly around the prey, and when the process is finished the stomach and its contents are sucked back into the starfish's body. The fluids then pass into the digestive pouches in the starfish's arms, where digestion is completed and the food absorbed.

Most shelled molluscs withdraw and close their shells when disturbed, but this does not protect them from the starfish. The latter wraps its arms around a closed bivalve shell and brings its tube-feet into play. The tube-feet are clamped onto the two valves and then the starfish contracts them in such a way that the valves are pulled in opposite directions. It has been estimated that the combined pull of the starfish's tube-feet and body muscles is equivalent to about 3kg (6·6lb), and this pull is able to overcome the strong closure muscles of a bivalve within five or ten minutes.

The slightest tiring of the closure muscles allows the starfish to separate the valves, and a gap of just 0·1mm is enough for it to start pouring in its digestive juices. It continues to pull, and before long the closure muscles give way and the starfish can push its whole stomach into the mollusc to digest it. It seems that the starfish does not get tired because it uses its tube-feet in relays and their muscles can rest for a while.

Starfishes are often major pests on oyster farms, where they open vast number of the molluscs in a very short time. The oyster farmers actually increased the damage in earlier times through their lack of knowledge of the starfish's amazing regeneration ability: they dredged up the pests, chopped them in pieces, and threw them back again, little dreaming that each piece was likely to grow the missing parts and become a complete starfish.

Independent feeding

We have seen that the tube-feet help starfishes to feed in several different ways, and recent research has shown that they can actually feed themselves as well. This is very important, because the water vascular system has very little connection with the rest of the body tissues and cannot share the normal food taken in through the mouth. The tube-feet absorb amino acids from the surrounding water and use them for their own nourishment. Most of the amino acids are probably absorbed while the starfishes are feeding in the normal way, for some broken-down material inevitably escapes from the everted stomach, but the tube-feet do not rely entirely on this source of material and they can absorb amino acids from ordinary seawater. It seems likely that the tube-feet of all echinoderms maintain themselves by this method.

The crown of thorns

Not all starfishes feed regularly on molluscs. The crown of thorns (*Acanthaster planci*) is a very large and spiny starfish of the Pacific region that eats corals. It settles on a coral colony, holds fast by means of its tube-feet, and everts it stomach over the polyps. These are digested and absorbed quite rapidly and the starfish glides to a fresh patch, leaving the first feeding area as a bare white space. An average-sized crown of thorns, about 40cm (16in) across, can denude as much as $1800cm^2$ ($280in^2$) of coral reef in one day.

There was a tremendous population explosion among the crown of thorns in the 1960s, when the animals were found at densities exceeding one per square metre on some parts of Australia's Great Barrier Reef. There are signs that the starfishes are declining again now, but tremendous damage was done to reefs in many places and it will be many years before the slow-growing corals recover completely.

No satisfactory explanation of the explosion has been put forward, but one contributory factor might be the excessive collecting of the triton (*Charonia triton*). This large sea snail, whose beautiful shell is much prized by collectors, is one of the few known predators of *Acanthaster*. Some attempts were made to control the starfish by breeding large numbers of tritons and releasing them on the reefs, but the results of these experiments were inconclusive.

Burrowing starfishes

A few starfishes have adopted a burrowing habit. These species usually have less distinct arms and their tube-feet, which generally lack suckers, are used for digging and for plastering the tunnel walls with strengthening mucus. Some of the burrowers feed on particles of debris, which they waft into their mouths with their tube-feet, but most remain predacious and feed on a variety of crustaceans, worms, and molluscs that they discover in the mud and sand.

Destroyer of the reef
Right The voracious crown of thorns starfish feeds on coral and has damaged large areas of Australia's Great Barrier Reef. The starfish simply everts it stomach over the coral and pours out digestive juices to destroy the living tissues. As it moves to the next patch it leaves bare, white coral rock where there was once colourful living coral.

Water Beetles and Dragonfly Nymphs

Aquatic beetles, like those living on the land, include both herbivorous and carnivorous species, but it is the carnivores that attract most attention, not least as a result of the damage that they inflict in garden ponds and jam-jar aquaria. Some of the fiercest carnivores belong to the Dytiscidae, a worldwide family with about 4,000 members. The best-known of these — in fact, probably the most studied of all beetles — is the great diving beetle (*Dytiscus marginalis*), which can be found throughout much of the northern hemisphere.

The great diving beetle

The adult *Dytiscus* spends much of its time hanging motionless just below the water surface or clinging to the water weeds, but it is always on the lookout for prey — worms, crustaceans, insect larvae, tadpoles, or even small frogs and fishes. The large eyes are very efficient at detecting prey, but the beetle also needs to pick up the scent of its prey

before being stirred to action. When it does attack, it swims rapidly forward with the aid of its broad-fringed hind legs and grabs the victim with the front two pairs of legs. The scissor-like action of the jaws then quickly carves up the prey and passes it into the mouth.

The water tiger

The larva of *Dytiscus marginalis* is such a powerful predator that it is known as the water tiger in the United States. It feeds on much the same types of prey as the adult, but its feeding mechanism is rather different. Its eyes are less efficient than those of the adult and it relies more on scent to detect its prey. It walks or swims towards the prey and stabs it with the immense, curved jaws.

The water tiger — a powerful predator
Below The fearsome larva of the diving beetle (*Dytiscus*) will attack prey much larger than itself, although here it has plunged its great jaws into a small insect. Food is sucked up through a slender canal in each jaw.

Each jaw contains a slender canal through which digestive juices are pumped into the prey. Digestion of the prey thus takes place outside the larva's body, and the liquefied food is pumped back up through the jaws and into the gut. The victim is reduced to an empty husk and discarded.

Dragonfly nymphs

Young dragonflies, like the adults, are entirely predatory creatures but, being aquatic, their diets and their methods of prey capture are obviously very different from those of the adults. Their food ranges from microscopic protozoans, through worms and insects — including other dragonfly nymphs — to tadpoles and even small fishes, but exactly what is taken depends on what is available and also on the size and species of the dragonfly nymph involved.

Detecting prey

In their earliest stages, when the nymphs are no more than a few millimetres long, they feed on protozoans and other minute creatures. These are detected by the antennae, but not until they are very close to the nymphs. When a nymph detects the presence of prey, probably through the minute disturbances set up in the water, it brings its antennae together and uses them as direction-finders to line up on the creature before attacking it.

The nymphs of most damselflies (see page 152) retain the basic feeding behaviour throughout their lives, with the antennae remaining the major organs of detection. They spend much of the daytime clinging to water plants or to debris on the bottom of the pond or stream and waiting for suitable prey, such as a water flea or a mosquito larva, to come within range. At night, however, especially if food has not been plentiful during the day, they may go searching for prey.

Although the antennae remain the most important sensory organs, tactile organs on the legs also play a part in detecting prey. The eyes are poorly developed, and probably cannot see movements more than about 3cm (1.2in) from the body, but they may play a small part in detecting food in the daytime.

A number of dragonfly nymphs, including those of *Orthetrum* and *Cordulegaster* species, bury themselves in mud and detritus and use their antennae, legs, and sense organs around the jaws to detect any suitable prey that may come along. Their eyes may also play some part, although they have a maximum range of only about 5cm (2in). Nymphs living in this way are among the slowest-growing dragonflies.

The most active dragonfly nymphs are those that crawl among the water weeds in mid-water and in the surface layers. They include many of the hawker dragonflies (see page 152) in the family Aeshnidae, and also the damselflies of the genus *Lestes*. The eyes are the main sense organs in these nymphs, being able to detect prey up to 20cm (8in) away. Even when the nymphs are only a few days old and still relying on their antennae to find food, the future importance of the eyes is apparent; most damselfly nymphs and bottom-living dragonfly nymphs have only seven ommatidia in each compound eye at this stage but *Lestes* nymphs have about 30 ommatidia and the aeshnids have between 150 and 300 in each eye.

By the fifth instar — when the nymph has changed its skin four times — the eyes are functioning well and most hunting is carried out by day. Having sighted suitable prey, the nymph moves purposefully towards it and generally achieves a capture. Some nocturnal hunting occurs, however, with the antennae and tactile organs on the legs detecting the prey as they do in other species.

The mask of death

Whatever method the nymph uses to detect its prey, the method of capture is always the same, involving an elaborate structure known as the mask. This is the much-modified lower lip, or labium, and it is unique to the dragonflies. It is called the mask because it conceals most of the other mouth-parts when the insect is at rest. There are three main regions to the mask: the postmentum, which is attached to the head; the prementum attached to the postmentum by a hinge joint; and the clawed or toothed palps at the free end. When at rest, the postmentum is folded back between the legs, with the prementum tucked up underneath it and the claws close to the jaws.

When prey has been detected and the nymph has positioned itself for the kill, the mask is fired out by hydraulic pressure. Contraction of muscles in the thorax and abdomen force body fluids forward into the mask and cause it to shoot forward with the claws agape. Muscular action then closes the claws and draws the mask back with the prey.

The broad-bodied nymphs of the anisopteran dragonflies (true dragonflies) possess a diaphragm across the abdomen, and this helps the firing of the mask by forcing all the fluids forward. It is thought that the very narrow abdomen of the damselfly nymph has the same effect without the need for a diaphragm.

As soon as the impaled prey has been retrieved, the horny jaws get to work to chew it up and shovel it into the mouth. Dragonfly nymphs can consume quite large prey at one sitting, but they can also go without food for considerable periods — very necessary for those species that bury themselves on the bottom and wait for food to arrive.

Water flea for dinner
Left The nymph of a damselfly (*Pyrrhosoma nymphula*) has impaled a water flea (*Daphnia*) on the claws of its mask and is beginning to eat it. This kind of nymph lives at the bottom of ponds and hunts largely with the aid of its antennae.

Death in the weed
Below The nymph of a large dragonfly crawls among the water plants in its search for food. Nymphs of this type have relatively large eyes and hunt by sight after the first few weeks of life. The mask is clearly visible.

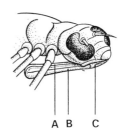

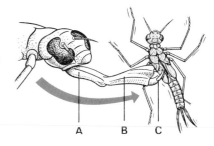

The mask in use
Left When at rest, the mask is folded below the head, with the prementum (B) under the postmentum (A).

Below When prey is detected, the mask is shot out and the prey is impaled on the powerful claws (C).

A   B   C

A   B   C

Penguins and Auks

We have seen earlier in this book (pages 166 to 169) that the majority of seabirds obtain their food either by snatching it from the water as they hover or fly over the surface or by diving right into the water from a considerable height. Now we are dealing with birds that employ a very different method – birds which float or swim on the surface and dive from there to seek and catch their prey. The two major groups of birds in this category are the penguins – which are flightless and thus bound to feed in this way – the cormorants, and the auks, which include such well known birds as the guillemots and the puffin. The divers, or loons, and various ducks also feed in this way, as described on page 182.

The penguins – cold water addicts

Penguins all inhabit the southern hemisphere but, contrary to popular opinion, they are not all denizens of the icy seas of Antarctica. The Galápagos penguin (*Spheniscus mendiculus*) reaches almost to the Equator, and several other species live around the coasts of South America, South Africa, and Australia. They are all associated with cold currents, however, for they rely on the cold waters to provide their food. The birds live entirely on marine animals, and these depend upon planktonic plants which themselves depend upon a rich supply of minerals. Since minerals are most abundant in areas where cold currents meet warm waters, this is where the greatest growths of plankton and hence the largest populations of penguins occur. Millions of the birds gather to breed on the coasts near these rich feeding grounds and collect thousands of tonnes of food annually.

Penguins feed almost entirely on fish, squid, and crustaceans, but each of the 18 species of penguin specializes to some extent. One investigation of the Magellanic, or jackass penguin (*Spheniscus magellanicus*), for example, revealed that 94 percent of its intake was fish – mainly anchovies and pilchards; 5 percent was squid; and only 1 percent was crustacean.

The emperor and king penguins, on the other hand, feed largely on squid, while the Adélie penguin and its relatives exist mainly on crustaceans. A total of 960 small crustaceans was found in the stomach of one gentoo penguin (*Pygoscelis papua*) – little wonder that the droppings of these birds are often bright pink. The crustaceans involved are mainly species of *Euphausia*, commonly known as krill. This is also the food of the giant whales, and it is interesting to note that many penguin populations have increased dramatically during the 20th century as the big whales have been hunted almost to extinction.

Specialization in feeding is by no

The squid-catching king penguin
Above The long, slender beak of the king penguin is ideally constructed to catch slippery squids. The bird also catches quite a lot of fish and krill. The king penguin lives on and around the sub-Antarctic islands, in water much warmer than those inhabited by its larger relative, the emperor penguin.

means complete, and a penguin that relies mainly on fish in one area may feed mainly on squid elsewhere. The rockhopper penguin (*Eudyptes crestatus*) exists on fish and crustaceans around Cape Horn, for example, but turns more to squid around Tristan da Cunha, where the krill is less abundant. The ability to utilize several sources of food enables penguins to colonize many different areas, and it also allows several different species to live in one area without too much competition. This is very important at breeding times, when vast amounts of food are taken from small areas of the sea.

Where closely related species, such as the rockhopper and the macaroni penguin (*Eudyptes chrysolophus*), live together they avoid competing for food because they are of slightly different sizes and thus take different-sized food. The larger penguins may also go further down for their food, and the breeding of the two species may be out of step so that they are not all gathering food for hungry chicks at once. Chinstrap penguins (*Pygoscelis antarctica*) often breed in association with Adélie penguins (*P. adeliae*), but the chinstraps feed closer to the shore than the Adélies and so the two species avoid direct competition.

Assorted beaks

The penguins are excellent swimmers and divers. They use their strong flippers to 'fly' underwater at speeds of up to 36kph (22mph), and they catch most of their food between 10–20m (33–66ft) down. Most dives last for no more than about three minutes, although the em-

Sociable razorbills
Right A group of razorbills perch on a cliff and display their sharp-edged fish-catching beaks. The birds feed on small fishes, such as sand eels, smelts, and sardines, carrying them crosswise in the beak like the puffin (far right).

Feeding the baby
Below An adult chinstrap penguin feeds its youngster with regurgitated food. Like the Adélie penguin and some others, the chinstrap regurgitates only as far as its throat, and the youngster pushes its head into the mouth.

peror penguin (*Aptenodytes forsteri*) has been known to dive for 100 minutes and to reach depths of 265m (869ft).

Food is detected by sight and caught with the only available tool – the beak. This varies a good deal in size and shape according to the main foods of the various species, although the specializations are not so great that the birds cannot turn to other foods. Adélies, and most other penguins that concentrate on krill, have short beaks and a weak biting action, and can deal only with small prey. Adélies rarely take anything longer than about 7·5cm (3in). Gentoo penguins feed mainly on krill, but they have much longer beaks and they can deal with larger prey than the Adélies. They regularly take fishes and squid up to 12·5cm (5in) in length.

The beaks of the emperor penguin and the king penguin (*Aptenodytes patagonica*) are very long and slender and, although these penguins do catch quite a lot of krill, they are best fitted for catching long and slender squid. The Peruvian penguin (*Spheniscus humboldti*) and other members of its genus are primarily fish-eaters

Beaks full of fish
Right and below The puffin dives from the surface and 'flies' through the water, often right to the sea-bed, to look for sand eels – slender fishes, but not related to true eels. By using its tongue to hold captured fishes against the toothed edges of the upper mandible, the bird can go on catching more, until it has as many as twenty fishes lying crosswise in its stout beak.

The pygmy cormorant
Below Cormorants have relatively poor water-proofing and often hang their wings out to dry.

and they have heavily-built beaks with strong muscles and a strong bite. In relation to their body size, they have a much wider gape than the other penguins, allowing them to deal with relatively stout fishes.

The tongue and palate are covered with spines in all penguins and clearly help to hold what is often very slippery prey. Specialist krill-feeders, such as the Adélie penguin, have exceptionally large tongues that rake the small prey to the back of the throat to be swallowed.

Penguin hunting methods

Most penguins hunt independently, although there may be hundreds fishing in a small area, but there are reports of cooperative fishing in some species. The blue penguin (*Eudyptula minor*) and the black-footed penguin (*Spheniscus demersus*) both sometimes work in groups to herd fishes into tight shoals so that they can be picked off with little effort.

The penguin swims forward with its head moving steadily from side to side as it looks for food, and the prey is snapped up with darting movements of the beak. Even krill is caught in this way, there being no evidence that penguins swim through shoals and scoop up mouthfuls at a time. Fishes are usually grabbed near the middle of the body, but with a few flicks of its head the penguin manages to turn the prey round and swallow it head first. This may happen under the water, or the penguin may surface to swallow its catch.

During the incubation of the eggs and the early lives of the chicks, the parent penguins take turns to go to sea to feed and collect food for the chicks. Food for the chicks is carried back in the stomach and regurgitated, but even this is not an entirely safe method, for the scavenging sheathbills often turn pirate and snatch up the regurgitated food before the penguin chicks get to it. Adélie penguins

avoid such interference because the chicks normally push their beaks right into the parent's throat to get the food.

The auks

The auks have been described, quite rightly, as the penguins of the north, for, although they can fly, they are remarkably like penguins in most of their habits. All have their legs far back on the body, and thus walk upright on land, and most use their wings to swim under the water.

Apart from the cormorants, which are found around nearly all the world's coasts and on many inland waters as well, the auks are confined to the northern hemisphere. Among the best known of the 21 species are the puffin (*Fratercula arctica*), the razorbill (*Alca torda*), the guillemots, or murres (*Uria* spp), and the little auk, or dovekie (*Alle alle*). The latter, which is one of the most abundant seabirds, is only about 20cm (8in) long and, like some of the smaller penguins, it feeds mainly on crustaceans. It even has the unusually large tongue that is characteristic of these penguins.

The larger auks – which are no bigger than the smallest penguins – prefer fish and squid, although they also eat a good deal of crustacean food. They generally catch their food at depths of about 10m (33ft). Around the coasts, many of them go right down to pluck sand eels from the bottom. The puffin is famous for its ability to carry twenty or more of these little fishes crosswise in its beak at once.

Most auk species breed in dense colonies on cliffs, often with several species nesting in close proximity. As among the penguins, this leads to a great drain on the sea's resources, but the auks parallel the penguins in their habits and each takes slightly different food or feeds at a different depth or at a different distance from the shore, thus ensuring that there is enough food for all.

The cormorants

Cormorants fish in both fresh and salt water, often ducking their heads under the water to look for prey. They dive after their prey and then bring it to the surface to swallow it, often beating it on the water to weaken it so that it can be turned and swallowed head first. Some cormorant flocks are thought to round up fishes like the penguins do, or to drive shoals into the shallows like pelicans.

Fishermen often hunt cormorants because they interfere with fishing operations and reduce the catch, but Japanese fishermen have turned the cormorant's habits to good use. The birds are actually trained to bring fishes back to the boats or to the shore. Leather collars around their necks prevent them from swallowing their catches.

Ducks, Divers and Darters

These three groups of water birds, together with the grebes, all feed at or from the water surface. They are all superficially duck-like, although none is closely related to any other and their feeding methods show a number of differences.

Fast swimming divers

The four species of divers, belonging to the family Gaviidae, all live in northern waters. They breed by fresh water, but normally spend the winter in coastal waters. Divers are supreme in the water, but very ungainly creatures on land. Their legs are set very far back and only the webbed feet actually protrude from the body. Lacking the upright stance of the penguins and the auks on land, the divers can only shuffle along and they do not usually come ashore other than to breed. In the water, the rear position of the 'engine' drives the birds along at high speed, both on the surface and below it. When diving, the birds propel themselves entirely with their feet and use their wings as stabilizers.

Divers do not generally stay under the water for more than about a minute, and most feeding dives last no more than 30 seconds, but a frightened bird can stay down for five minutes and swim several hundred metres during this time to escape an enemy. It relies on reserves of oxygen stored in its breast muscles during these long dives. Reports of dives lasting for 15 minutes and covering several kilometres are almost certainly based on inaccurate observations; the birds are able to lower their bodies into the water and swim with just the top of the head and the beak above water, and in such a position they can easily be 'lost' in all but the calmest water, giving an observer the impression that the birds have completely submerged.

The diver has extremely good eyesight and relies entirely on it to find fishes and other food, such as crayfishes and other crustaceans, under the water. Its speed then enables it to overhaul and catch the prey in its sharp, stout beak. Fishermen sometimes complain that divers take a lot of trout, but this seems unlikely because the birds maintain quite large territories when they are on fresh water in the breeding season and they would not have much effect on fish stocks. They could reduce fish populations in certain coastal waters during the winter, however, for divers occasionally form small flocks at this time of year and feed gregariously.

The grebes

The grebes of the family Podicipitidae are similar to the divers in many of their habits, although they have longer and

more slender necks and smaller bodies. They also have lobed toes instead of webbed feet but, as in the divers, the feet are set far back on the body and they work almost like propellers to push the birds efficiently through the water. Fishes, amphibians, crustaceans, insects, and certain amounts of vegetable matter make up the grebe's diet, most of the animal food being taken under the water after a fast chase.

In rough water, the birds may swim with their heads bent down into the water so that they can see prey more easily, but they normally hunt on the surface and sink quietly down into the water when they see something. Like the divers, mergansers, and darters, the grebes can alter their specific gravity, or buoyancy, by compressing their feathers and driving out some of the trapped air so that they gradually sink into the water. Prey is caught in the beak and brought to the surface before it is swallowed. Fishes often have to be juggled about before they can be gulped down head-first.

Grebes have a very strange habit of plucking out some of their own body feathers and swallowing them. The feathers then accumulate in the stomach and it is thought that they help the birds to deal with fish bones. The bones become bound up with the matted feathers and this prevents them from damaging the stomach wall. They may remain in the stomach until they are completely digested, or the whole feather and bone mass

may be regurgitated as a pellet. The darters, or anhingas, (see below) have a furry lining to the stomach that may act in a similar way.

The ducks

The duck family (Anatidae) is a very large one, containing the swans and geese as well as the ducks themselves. They exhibit a wide range of feeding habits, from purely vegetarian to exclusively carnivorous, and there are corresponding variations in the form of their beaks. The geese, which feed by grazing on land, have sharp-edged beaks that cut off the grass like scissors. Most of the ducks have filtering devices along each edge of the beak, formed by a number of ridges and furrows that interlock when the beak is closed. The duck takes a mouthful of mud or muddy water, closes its beak, and then forces the water out

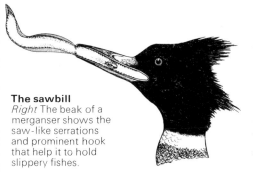

The sawbill
Right The beak of a merganser shows the saw-like serrations and prominent hook that help it to hold slippery fishes.

The beak that spears
Right Most predatory water birds catch prey *in* their beaks, but the darter, or anhinga, seen here actually uses its beak as a spear and sticks it into the prey. The bird then has the problem of getting its beak out again so that it can swallow the food. Adult birds often do this with a deft flick of the head and then catch the fish cleanly.

through the filter. Food particles, together with much inedible material, are retained on the ridges, or lamellae, and then swept into the throat by the tongue. The arrangement of the lamellae depends upon the preferred diet of the bird. Mallard and other ducks that sift plant fragments and small animals from the mud have relatively coarse filters, but the shoveller (*Anas clypeata*), which is a plankton-feeder, has very fine lamellae. It swims along with its broad beak half-submerged and gathers food from the surface layers of the water.

The sawbills, such as the goosander (*Mergus merganser*), are fish-eating ducks in which the margins of the beak are strongly serrated to form large, backward-pointing teeth. These are ideal for catching and holding slippery fishes. There is also a prominent hook at the front of the beak, which helps with the initial capture. Sawbills are very similar in general appearance and habits to the cormorants (see page 180), but the similarities are purely the result of convergent evolution (see page 116) and do not imply a relationship between the two groups. Sawbills feed in both fresh and salt water and catch their prey by swimming very rapidly under the surface. They are aided in this by the posterior position of their large, webbed feet. Their ability to submerge without a ripple also enables them to take fish by surprise. All kinds of fish are eaten, including pike and eels, and the sawbills also catch frogs and

crayfish when they are in fresh water.

The eider duck (*Somateria mollissima*) keeps almost entirely to coastal waters and feeds by diving to the seabed for cockles, mussels, crabs, and other relatively large invertebrates, including starfishes. It shares these habits with quite a number of other large sea ducks, such as the scoters (*Melanitta* spp), but competition is avoided because each species has its preferred depth for feeding. The eider, for example, usually feeds in water 3–6m (10–20ft) deep, whereas the scoter feeds further from the shore in water depths of about 18m (60ft). The beak lamellae are not very marked, for most of the food is picked up individually and little sieving is necessary. Prey is generally swallowed whole, and the powerful gizzard grinds it up. Even the thick shells of whelks and cockles can be dealt with by the gizzard. The eider has a special way of dealing with crabs, however: it grasps them by two or three of the legs and jerks them about on the surface until the legs break off; these legs are swallowed and the crab is rapidly caught again so that the process can be repeated until all the legs and pincers have been removed; the body is then swallowed.

The snake-bird, or darter
The four species of darters live in fresh waters in America, Africa, southern Asia and Australasia. All belong to the genus *Anhinga* and they are closely related to the cormorants. They are

called snake-birds because of their incredibly long necks and slender heads, but in America they are generally referred to as anhingas. They swim strongly both on the surface and underneath it, and, like the divers, they can also swim with just the head above water. Prey is sought by swimming under the water and consists of fishes, frogs, crayfishes, and various other animals. Small prey may be picked up in the beak in the normal way, but most food is speared.

When the darter is chasing prey under the water it may use both wings and feet to swim, and its neck is doubled back with the beak lying close to the breast. As the bird approaches its prey, the head moves backwards and forwards rather in the way that a dart player moves his arm a few times before releasing the dart. It may be that this movement enables the bird to assess the range with its eyes, and when the range is right the neck streaks out and the beak strikes into the prey like a harpoon. The bird then surfaces and begins the somewhat tricky task of removing the catch from the beak. This is sometimes accomplished by a quick flick of the head, which throws the prey into the air so that it can be caught and swallowed head-first. Alternatively, the bird may hit the prey against a branch to loosen it. Young birds have more difficulty in removing their prey than older ones, suggesting that, although instinctive, the procedure needs some practice to get it right.

The Amazing Electric Fishes

The small electrical discharge that occurs when a nerve impulse reaches a muscle is essential for the life of almost every animal, but only in fishes have special electric organs evolved in which these standard discharge levels have been greatly amplified. By elaborating the system of producing and detecting electrical disturbances, electric fishes have acquired an extra sense that can be used to aid navigation and as a means of feeding and defence. The electric fishes belong to six families, and the evolution of their special organs is a striking example of convergent evolution; in each case the organs are derived by the modification of a different type of muscle to enable it to produce electrical discharges of various strengths.

How electric organs work

The basic units of all these 'batteries' are small flattened cells called electroplates, each of which may simply be a single transformed muscle fibre. Each electric cell is sheathed from its fellows and they are stacked in columns rather like piles of coins; the configuration of the columns varying between the different fish groups. The electric potential of each electroplate is about that expected at the junction of a nerve and a muscle, approximately 150 millivolts, but by adding together the discharges of many cells connected in parallel or in series a very much larger current or voltage can be produced. The plates are usually connected in series whilst the columns they compose are joined in parallel. It is usual for a freshwater fish to have fewer but longer columns of electroplates than a marine fish, to produce the higher voltages necessary to overcome the poor conductivity of fresh water. The electrical discharge can be produced either as a result of an external stimulus or at the will of the fish.

Navigation by electricity

One type of electric organ is that seen in the Mormyridae and species of *Gymnarchus* of tropical Africa, where it is the tail muscles that have become modified. These are fish-eating species of murky fresh water, and their canny powers of prey detection and avoidance of obstacles cannot be attributed to their sight, which is poor, even when it is used together with the lateral line. Experiments have revealed that these fishes produce an oscillating electric field about themselves, which they maintain by irregular low voltage pulses from the electric organs in the tail. They swim by rippling the dorsal fin while the tail is held stiff so that the axis of the body remains in line with that of the electric field. Any object whose conductivity differs from that of the water is detected

The most powerful living generator
Above The electric eel from South America has enormous 'batteries' accounting for perhaps half its weight. It can produce discharges of up to 650 volts — enough to kill a horse.

Food-finding with an electric field
Below Gymnarchus (left) and the skate surround themselves with electric fields generated in the tail. Disturbance of the field by prey is immediately detected by the head.

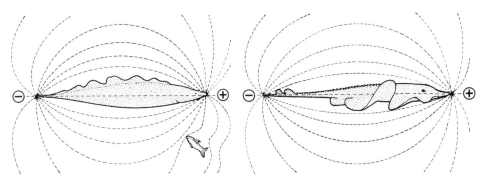

The knife-fish
Left The South American knife-fish (*Sternarchus albifrons*) uses its electric field for both navigation and food-finding. It keeps its body rigid and swims by waving the long ventral fin.

An electric jacket
Right The electric catfish (*Malapterurus electricus*) has its 'batteries' in the form of a jacket wrapped around the trunk. They give out strong discharges whenever the fish is touched.

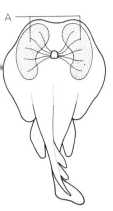

The electric ray
Left Electric rays have powerful 'batteries' between the head and the broad fins. The drawing shows the large nerves that charge the batteries (A).
Right The fish normally conceals itself in the sand, and when it is touched it emits a discharge powerful enough to stun prey. The fish can also leap up and wrap itself around its prey.

as it enters this field because it disturbs the electrical pattern. Minute electrical distortions of only a few millivolts can be perceived by tiny pits in the skin, which are connected by small canals and linked to a specially enlarged portion of the hind brain. There are over 120 species of these mormyrid fish and their electric organs have certainly contributed to their success in a rather inhospitable environment.

The South American knife-fishes, such as *Gymnotus*, are not closely related to the Mormyridae, but their electric organs, in this case derived from trunk muscles, have a similar function. The body is similarly held stiffly while the long anal fin propels the fish, and the electric sense is probably used for navigation and prey detection in just the same way as it is in the African species.

Electricity as a weapon

The South American electric eels, such as *Electrophorus*, are included in the knife-fish family (Gymnotidae). Their electric organs may constitute half the mass of the body. Three separate organs are present, with one very much larger than the others. The body, which can reach 3m (10ft) in length, is held stiffly during swimming and the two smaller electric organs pulse to produce a low-voltage perception field. This extra sense is a great advantage to these fish because they have very small eyes and live in turbid water. The larger electric organ possesses about 120 columns of between

6000 and 10,000 electroplates each that produce discharges up to 650 volts. Such discharges have been known to kill horses, but they are more usually used to stun fish and frog prey.

The pulses are very brief, perhaps of only 2–3 thousandths of a second, but a series of several pulses can stun prey over quite a considerable range. The electric organ is long, and to ensure that there is a single simultaneous discharge the electroplates of the front end (where the discharge originates) have slower-acting nerves than those of the hind end. Contact with an electric eel will trigger off a spontaneous defensive pulse, which is in addition to discharges used for navigation and stunning prey.

The electric catfish (*Malapterurus*) of the Nile has another type of electric organ, which takes the form of an electric jacket about the trunk. These fish may reach about 1m (39in) in length and they feed voraciously on other fish species and invertebrates. The electric organ spontaneously discharges when the fish is touched and this defensive role is probably the organ's most important role, although some captive specimens have been seen to stun prey with such 350-volt discharges.

The electric rays (Torpedinidae) of temperate and tropical seas have evolved their electric organs from modified gill muscle, and they appear as large kidney shapes on either side of the head, supplied by cranial nerves. The columns are stacked vertically and in *Torpedo torpedo* number about 2000, each containing some 1000 hexagonal electroplates; this is a moderately sized species and its 'battery' can emit currents of up to 50 amps at 60 volts. The largest species, *T. nobiliana*, may deliver a shock of nearly 200 volts. When prey is sighted, the torpedo pounces, wrapping around the pectoral fins and then discharging the electric organ. The effectiveness of this feeding method is demonstrated by the fact that these sluggish, bottom-dwelling rays often prey upon very swift fish, such as salmon, which could not possibly be caught without the use of electric organs to stun them from a distance.

The conspicuous spots on the back of most torpedoes are an example of warning coloration, indicating that predators should leave them well alone. This may be particularly protective to a fish that has just fed because the electric organ needs time to recharge before it can be used again. The fishermen that sometimes catch electric rays certainly heed the warning of their markings, the price for ignoring them being painful and often long-lasting shocks. Certain small torpedoes (*Narcine*) have smaller secondary electric organs that may function as navigation aids, but this is only a hypothesis at the moment.

The lure-feeding stargazers, such as *Astroscopus*, are another marine group bearing electric organs, in this case formed from modified eye muscles. These are bottom-dwelling fish that lie buried in sand with only their eyes and gaping mouths exposed, and the quite considerable current produced from the relatively small electric organs may serve to protect the vulnerable eyes or to stun prey as they approach the lure, which is attached to the floor of the mouth.

The last group of electric fish are the skates and rays of the family Rajidae, which have very small, weak electric organs in the tail. These have only ever been seen to discharge weak shocks under extreme experimental stimulation and biologists have so far been unable to ascribe any function to them in nature, other than that of navigation.

Convergent evolution

Since these six groups of fish are not closely related, occur both in fresh and salt water, and use different parts of the body to generate electricity, it is clear that electric organs, despite their complexity, have been evolved at least six times. Even those fish that do not produce high voltages to stun prey may still gain considerable advantages from using low-voltage discharges to aid navigation. In this case the electric sense acts as a kind of 'radar,' enabling the fish to be active at night and in murky water; a considerable advantage over hunters that must rely entirely on sight.

Ferocious Freshwater Fishes

A number of highly predatory fishes are found in freshwater throughout the world. They lack the sustained speeds found in many of the marine predators, but they are usually able to rely on a certain amount of concealment among the water weeds. They can thus stalk their prey, or even lie in ambush and dash out to catch prey as it passes. This habit is epitomized by the pike found in freshwaters of the northern hemisphere.

Pike

Pike belong to the family Esocidae and five species of *Esox* are recognized, ranging from the small (35cm/14in) chain pickerel of the eastern United States to the muskellinge of the Great Lakes, which may exceed 2m (6·5ft) in length and some 50kg (110lb) in weight. The familiar northern pike (*Esox lucius*) of Europe is typical of the family in its design as a first class hunter. The body is long and torpedo-shaped with a big bony head bearing prominent eyes set high up on the sides looking forwards and upwards. The snout is broad and flattened, and scored with a number of grooves running from the eyes to its tip. The mouth is large, bristling with fierce teeth, those of the lower jaw being straight and needle-shaped while the smaller, and more numerous, teeth of the upper jaw are backwardly pointed. The body is perfectly camouflaged amongst water vegetation with olive-coloured upper flanks barred and spotted in buff. The tail is broad and powerful and together with the dorsal and anal fins, which are set well back down the body, produces short bursts of very fast swimming.

The solitary pike lies in ambush concealed amid weed, but because of the buoyancy of the swimbladder it can hover amongst the plants in mid-water, needing only to scull the pelvic and pectoral fins occasionally for orientation. The prey is detected by eye and the relatively large size of the optic lobes of the otherwise small brain demonstrates the importance of these organs, which can spot fish moving as much as 15m (49ft) away in well-lit water. Pike experimentally blinded can still feed successfully, which suggests that the sensitivity of the lateral line system may also be used in the capture of prey that set up vibrations in the water in the vicinity of the pike. The pike lurks posed to strike and sights the prey along the grooves of the snout. Suddenly, and with great speed, the lunge is made, over perhaps as much as 10m (33ft), and the prey is seized, unable to escape from the curved teeth. Usually, fish prey are swallowed head-first and they may need to be turned round in the jaws before this is possible.

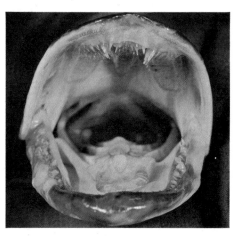

A snack for a pike
Above A young pike has caught a ten-spined stickleback, and must turn it round before swallowing it to avoid harm from the spines.

Not-so-tender trap
Below The prey's eye-view of an approaching pike, showing the formidable array of sharp teeth, from which escape is unlikely.

Digestion begins by the action of strong acids secreted onto the prey as it is swallowed, but the process is slow and a pike may lie quietly for several days with a distended stomach, not feeding again until digestion is completed. Young pike feed on worms, zooplankton and fish fry, but they grow quickly and are soon capable of taking fish of their own size and often of their own kind. Bigger pike eat large quantities of fish but voles, frogs, and water birds may also be consumed. A pike will generally attempt to eat the largest prey available to it, simply for reasons of energy economy; clearly it would be wasteful to lunge repeatedly at small fish when a single larger individual would provide the same amount of food for less effort. When times are hard, however, even large pike may resort to eating worms and water beetles.

Pike have earned a reputation for greed, no doubt due to the common sight of an individual killed by asphyxia, with the head of a large prey stuck in its throat and preventing adequate aeration of the gills. The maximum size of the northern pike and also its feeding capabilities are surrounded in folklore. Certainly specimens of 20kg (44lb) have been caught in Britain; and housed in a Scottish museum are the remains of a pike skull believed to be from a monster of some 32kg (70·4lb) taken from Loch Ken during the Jacobite rebellion of the 1740s. Clearly such a fish, reputed to have measured over 2m (6·5ft) long and possessing 20cm (8in) jaws adorned with wicked teeth, could quite easily have performed the dog and swan-swallowing feats described in country tales. The best tale is told by Gesner in 1558 of the so called 'emperor's pike', which, when caught in 1497, weighed 250kg (550lb), was nearly 6m (20ft) long, and bore a tag showing it to be 260 years old!

Garpike

The freshwater gars (*Lepidosteus* spp) of North and Central America show several primitive features, being the surviving members of a very old family that flourished many millions of years ago during the Mesozoic era. They resemble the pike in shape, being long and slender with median fins set well back towards the tail, but their snout is drawn out into a long slender beak studded with many small short teeth. These fish have a stiff body that is rather inflexible due to a covering of an ancient type of bony scales that do not overlap like the scales of most fishes. Gars are well camouflaged to lie concealed amongst weed, and often resemble pieces of drifting log. Although, like the pike, gars appear lethargic, they, too, are capable of short bursts of fast swimming. The eyes are very prominent and they hunt primarily

Ferocious piranha
Left Piranhas hunt primarily by scent, collecting the scent particles in the large nasal pits, which can be seen quite easily in this specimen of *Serrasalmus nattereri*.

Razor-sharp teeth
Right The mouth of a Brazilian piranha reveals the incredibly sharp teeth with which these fishes carve chunks of flesh from their victims' bodies.

Long-nosed garpike
Below The long snout of the garpike is full of many small, sharp teeth. The animal hunts with the aid of its large eyes and snaps up smaller fishes with a rapid sideways lunge.

by sight, and once close enough to the prey they sieze it suddenly with a sideways slash of the wicked snout. Gars are mainly fish-eaters and, having swallowed the prey head-first, digestion proceeds fairly slowly. It is surprising, therefore, that they have a rapid rate of growth; they can reach 50cm (20in) in a year and some species reach 3·5m (11·5ft) when adult.

Piranha

Perhaps the best known and most feared of freshwater predatory fish are the piranhas of the family Characinidae. The accurately documented accounts of their extreme ferocity, together with wildly exaggerated tales brought back by travellers have earned these fish a dreaded reputation. The term piranha loosely describes about 18 species, but possibly the most ferocious of these killers are the species of *Serrasalmus*, which inhabit the rivers of Central and South America; they are particularly

common in Brazil. The individuals are often no more than 30cm (8in) long, and the largest, the piraya, attains only 60cm (24in); but the nature of their feeding makes even the smaller species truly effective predators.

The body is deep, with a keel along the dorsal and ventral surfaces, and is flattened laterally, ending in a rather blunt snout. The tail is slender but very muscular and with a broad fin to drive the fish forwards with great force. The olfactory portions of the brain are well developed and the external openings of the nasal pits are large and situated prominently above and in front of the eyes, all indicating that the sense of smell is of great importance. The eyes are also quite big and forward facing. The massive and extremely muscular lower jaw bears terrifying blade-like, triangular teeth. The edges are razor sharp for slicing and the points are able to pierce animal flesh with ease. The smaller teeth

of the upper jaw are equally lethal, and when the mouth shuts they fit exactly between those of the lower jaw.

Piranha swim in shoals of many hundreds and once blood is released from a prey at a kill the smell may attract other shoals, so even large carcases are quickly cleaned of flesh. Small fish form the staple diet, but cannibalism is also frequent and, in fact, any animal entering the water may be attacked. No doubt, any wounded animal emitting blood, even in very low concentration, will be detected by the olfactory apparatus of these fish, but piranha also seem very curious about unusual movements and perhaps also detect prey visually and by sensing vibrations through the water. The sheer strength of the jaws and cutting edge of the teeth make short work of any animal preyed upon and a capybara, a large rodent, weighing about 40kg (88lb) was savagely bitten to pieces within the space of a minute, leaving only the bare bones. Even the strength of a full grown bull cannot overcome a piranha attack and the animal is quickly weakened by its own struggling and loss of blood.

There are many reports of attacks on humans by these fish, and bathers or people crossing streams have often been badly injured, sometimes even killed. Indeed there is the story of a traveller on horseback fording a stream who was brought down by piranha and both man and beast were later found reduced to skeletons, although the clothes of the man were scarcely damaged. Piranha are probably most ferocious when the males are guarding the eggs during the rainy season and this might explain why one particular stretch of river can be very dangerous to cross while another nearby, in which perhaps no eggs have been laid, can be safe for bathing.

Crocodiles – Predators from the Past

The crocodiles and their relatives – the alligators and caimans and the gavial – are the nearest living relations of the great dinosaurs. There are about 25 living species, making up the order of reptiles known as the Crocodilia. The adults range from about 1m (39in) to 7m (23ft) in length. All live in the warmer parts of the world, usually in rivers and lakes, although a few species prefer estuaries and other coastal waters. There are slight differences in the biology and habits of the various species, but all are fiercely carnivorous and their feeding methods are similar. The following paragraphs apply to all the larger crocodilians apart from the fish-eating gavial.

Changing menus

Very young crocodilians spend much of their time on land, where they roam through damp, waterside vegetation and catch a wide range of insects, spiders, and molluscs. As they get larger – up to a length of about 1·5m (5ft) – they concentrate more on frogs and toads. All of these small prey animals are snapped up with a sideways flick of the snout.

Fishes enter the diet when the crocodilians are about one metre long and begin to feed in the water more frequently, and they provide the bulk of the food during the middle life period of the reptiles. Large water snails form an important part of the diet in certain areas, notably the Bangweulu Swamp region of Zambia, where even large Nile crocodiles exist mainly on these molluscs.

Most well-grown crocodilians, however, turn their attention to mammals, birds, lizards, snakes, and other members of their own kind. The Nile crocodile of Africa takes far more large mammals than the American alligator, but this is more a measure of the availability of large mammalian prey than of the reptiles' preferences.

At all times the crocodilians are great scavengers, but the larger individuals are particularly fond of searching out carrion, both in and out of the water. It has been said that the crocodiles and alligators do for the rivers what the hyenas and vultures do for the land.

The catch

Small invertebrates living among the waterside vegetation do not give the crocodilians much trouble. The prey is spotted by one or other of the eyes, and then, with a surprisingly fast sideways lunge of the snout, the crocodile snaps it up in its toothy jaws. Sight is the major sense involved in feeding, although taste is also important in the water, where scent particles from the prey are carried to the taste buds on the predator's tongue. Smell is also important on land.

Watchful sunbather
Above A Nile crocodile sunbathes, its mouth open to keep cool. Although resting, the animal is alert for food.

A toothy trap
Below An Indian gavial catches fishes in its narrow jaws. At times it sweeps them towards the mouth with its tail.

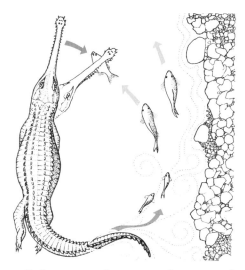

Fishes are often actively pursued through the water and caught with a sideways movement of the open jaws. They may also be caught with the aid of the powerful tail, which can flick them forward, often through the air, and towards the waiting jaws. Having caught a fish, the crocodilian surfaces and manoeuvres the prey with its teeth until it can be swallowed head first. Large fishes, such as the various kinds of catfish, may be beaten against rocks before being swallowed.

As the crocodilians get larger, they find it more difficult to chase fishes and they frequently turn to ambushing techniques, waiting for prey to swim within reach of the great jaws But the ambush is not set just anywhere, for the crocodiles are very good at picking out the best fishing grounds. At the beginning of the dry season in parts of Africa they often station themselves where small tributary streams join the main rivers. Large numbers of fishes come down the tributaries as they begin to dry up, and the crocodiles make some easy killings.

Water birds are generally pulled down as they float on or wade in the water. Swimming rodents are dealt with in the same way, but larger mammals are usually attacked when they come to the water to drink. People generally fall prey to crocodilians when they go down to the banks to bathe or to collect water – especially if they make regular visits to one place, for the crocodilians are quick to recognize such behaviour and they station themselves accordingly.

Animals may be knocked into the water by a blow from the reptile's tail, or, more usually, they are caught by one leg or by the muzzle and dragged into the water. The end result is usually the same – the prey is held under the water until it drowns. To ensure that it is heavy enough to sink in the water, a large crocodile may swallow several kilogrammes of stones, travelling considerable distances to collect them. It carefully regulates this 'ballast' to about

The ambush
Left Crocodiles obtain a lot of their food in the form of animals that come down to the water to drink. The reptiles often take up position close to favoured drinking spots, hanging just under the surface with only eyes and nostrils visible. As the antelope begin to drink, the crocodile submerges and swims slowly towards them. A sudden dart forward allows it to grab a foot or a muzzle in its powerful jaws, and it slides back into the water, holding its prey under until it drowns.

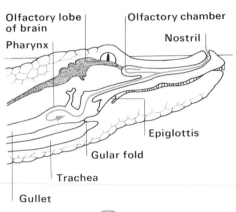

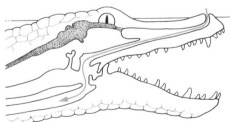

Sniffing crocodile
Left The crocodile has a keen sense of smell, but its sniffing is quite independent of its breathing: a flap known as the gular fold shuts off the windpipe when the animal is merely sniffing the air. *Bottom* As its lips are not watertight, the crocodile has to close its throat with the epiglottis when it breathes half-submerged.

Red in tooth
Right A crocodile feasts on a wildebeeste carcase. Its teeth are well suited for grabbing and ripping flesh, although they cannot chew. Teeth are continually falling out but, except in the oldest crocodilians, they are rapidly replaced by new teeth growing up alongside them.

one percent of its total body weight.

Nile crocodiles take mainly antelopes, zebra, buffaloes, and domestic sheep and cattle. They also take young hippos and rhinos in some areas. Even lions are not safe, for, although they can master crocodiles on land, they are easy prey for the aquatic monsters once they are floundering in the water after the initial attack. But the crocodiles do occasionally get too ambitious; they have been trampled to death trying to attack elephants, and bitten in half by the huge tusk-like teeth of hippopotamuses.

Torn to pieces
The crocodilian's teeth are all dagger-like and ideal for grabbing and holding their prey, but they are useless for cutting and chewing flesh. Fishes and other small prey are bolted down whole, but larger animals have to be dismembered before they can be swallowed. The belly can be ripped open and the soft parts torn out by taking a firm grip with the teeth and then shaking the prey vigorously. Part of the prey may be held down by one foot while the jaws tear at the flesh. Limbs are often torn off by a

'methodical' twisting; the reptile grasps the limb with its interlocking teeth and then twists its own body round and round with a vigorous thrashing action until the limb is torn from its socket.

Large prey cannot be consumed at a single sitting and, like the leopard, the crocodilian may well hide its kill for another day. The meat may well start to rot before it is eaten, but this does not worry the crocodilian. It was once thought that the prey was always stored and not eaten until it had started to rot and was thus easier to dismember.

The Playful Otters

Otters are carnivores that spend their time hunting along streams, in lakes or lochs, and along estuaries and seashores. Regrettably, destruction of their habitats, pollution, persecution, and hunting for their fur have made these beautiful creatures shy and more and more difficult to observe in the wild. They are one of the most joyful groups of animals to watch as they undulate their long sleek bodies through the water, twisting, turning and rolling in exploration and play. On land they bounce along on their strong, stubby legs, with the back arched high in the air, uttering excited shrieks, squeals, whistles and churring noises.

Otters live a solitary existence, the large males holding a territory of about 15km (9 miles) diameter, depending on the topography of the terrain and the availability of food. Each territory is well marked on prominent points along the banks by dung and by anal secretions known as spraints. The breeding females occupy smaller areas within the male territories. Subordinate nomadic males and females pass through these territories avoiding the owners, searching for a suitable area to set up on their own.

Webbed-footed fishers

There are twelve species of webbed-footed, river otters with the generic name *Lutra*. They are widespread, living in rivers, lakes and coastal waters of North and South America, most of Africa, Europe and Asia. They have dense, brown, waterproof fur, paler on the cheeks, throat and belly. The head is flat with small ears, nostrils that can be closed during a dive and long tactile vibrissae sprouting from the eyebrows and muzzle. The body is long and sinuous, ending in a thick, muscular tapering tail. The short legs end in five strong webbed toes, each with a sharp claw.

All are active fish-hunters. Lazily dog-paddling along the surface of the water, they dip their head down now and again to submerge their eyes and see if any prey is near. Their eyesight is as accurate under water as in air and plays a major part in locating prey. The eye is accommodated for underwater vision by rounding the lens and bulging out its front surface. This is accomplished with the help of the exceptionally strong sphincter muscles of the iris, which contract round the front portion of the lens to change its shape. On sighting its main prey, fish, the otter goes into top gear, diving and speedily propelling its way down through the water by flexing its spine and kicking in unison with its spread webbed hind feet. This provides the main forward thrust. The webbed front feet act as rudders to control the twists and turns as the predator follows every move of its fleeing prey.

As the otter swims along the surface, the long tactile vibrissae trail in the water and may alert the animal to any turbulence under the water. This tactile sense plays an important part in the location of prey in murky waters and the hairs can be used to feel for crustaceans, molluscs and frogs on the muddy river bed. Otters sometimes churn up the mud with their paws to disturb likely prey.

Although all the web-footed otters feed mainly on fish, other creatures are eaten when the opportunity arises and the variation in their diets reflects their habitat and the seasonal availability of the prey. The fish eaten by otters are those species that can be easily caught under river banks or in shallow water. Otters in temperate climates will take

The Canadian otter
Left A Canadian otter munches on a freshly-caught fish. Otters do most of their fishing in shallow water.

The clawless otter
Above Although it eats some fish, the clawless otter generally uses its fingers to dig out crabs and molluscs.

crayfish in the summer when they become active and accessible. Frogs, on the other hand, are more easily caught in winter, when they crouch torpid on the muddy river bed. They are also eaten in large numbers when they gather in ponds and ditches to spawn. Molluscs, reptiles, small mammals, eggs, nestlings and birds are also taken. Crabs and mussels as well as fish are eaten by otters living in coastal waters and hunting along the seashore.

Usually otters only eat fresh flesh and make no attempt to cache their kills, but otters hunting along the seashore have been observed hiding the remains of large fish under seaweed out of sight of scavenging gulls. Otters hunt alone or in family groups consisting of a mother and her cubs. A Canadian otter, *Lutra canadensis*, and her half-grown cubs have been seen hunting co-operatively, driving

fish into the shallows where they were easily caught. Similar observations are reported for small groups of giant otters. These large otters, with their massive webbed feet and powerful, flattened, paddle-like tails, live in sluggish streams in South America.

Sensitive feeders

Other otters have developed greater manual dexterity in their forepaws. They use their sensitive fingers to feel for crustaceans, molluscs, frogs and small fish in crevices and muddy waters. The webbing between the toes on their forepaws is reduced to just the base of their long mobile fingers and their claws are small or absent.

The little oriental small-clawed otter, *Amblonyx cinerea*, occurs in hillside streams in southern India, but elsewhere it is found in paddyfields, estuaries and coastal waters. This is the otter most commonly kept as a pet and seen in zoos. These delightful creatures spend much of their time in captivity playing with pebbles, feeling them, tossing and catching them in their hand-like forepaws. They often carry small objects in one hand, pressing it against the side of the body and gambolling along on three legs.

The African small-clawed otter, *Paraonyx* sp, is restricted to small torrential mountain streams in dense rainforests. They have short facial vibrissae and weak, shearing cheek-teeth and are thought to be more terrestrial than most otters. They eat frogs and soft-bodied land animals.

The African clawless otter, *Aonyx capensis*, has small claws on only the third and fourth toes of its hind feet, probably used in grooming the fur. The rest of its toes have small flat nails. These otters inhabit coastal plains, semi-arid country and dense rainforests, but they never venture far from sluggish streams or quiet pools. They feed on crabs and molluscs, using their large, flat cheek-teeth to crack open the hard shells of their well-protected prey.

The aquatic tool-user

The most enchanting of otters is the sea otter, *Enhydra lutris*, which is well known for its imitation of an oil-slick as it floats sedately on its back, bobbing up and down on the waves, clasping its new born young or its next meal to its belly. It lives among the kelp beds off the Californian shore and northwards along the western coast of North America, spreading across to the eastern shores of Kamchatka in Russia and to the Aleutian, Kurile and Commander Islands. Kelp does not have to be present in its habitat but the sea otter prefers rocky coasts and offshore islands. It is the most aquatic of all the otters, swimming in coastal waters up to 30m (98ft) deep; it seldom ventures more than a few metres on shore.

Its diet depends on what food is readily available in the habitat. The Californian otters dive for abalones, sea-urchins, mussels and crabs. The abalones are the major food item, bringing the otter into conflict with the local abalone fishermen. Like the clawless otters, the sea otter is very dexterous and will even use a stone as a hammer to dislodge its molluscan prey from the rocks. Floating on the surface, it places a stone on its belly and uses it as an anvil on which to smash the tough shells of the molluscs or brush off the sharp spines of sea-urchins. Young otters, with their weaker jaws and teeth, use stones more frequently than older animals, who can crack the hard armour with their teeth.

Females have been seen to dive for periods of one minute and males for slightly longer, gathering food with their right hand and placing it under their left armpit. There are reports of alarmed otters remaining under water for as long as four minutes. The northern sea otters, having no abalone beds in their habitat, feed mainly on sea-urchins, crustaceans, mussels, octopus and any fish they can catch. In some areas, fish makes up 50 percent of the sea otter's diet.

Eyes down
Above and right Dog-paddling at the surface, the otter puts its head under from time to time; if it sees a fish it dives after it.

A leisurely lunch
Below The sea otter enjoys a crab while floating lazily on its back. The 'anvil' is clearly seen.

Seals and Seal Lions

The seals are aquatic carnivores that, with a few exceptions, spend their entire lives in and around the sea. They come ashore to breed, and they often haul themselves out to sunbathe or sleep on the shore, but they get all their food from the water. There are three quite distinct groups: the true seals, the eared seals, and the walrus. The true seals (family Phocidae) have no obvious ear flaps and they are unable to bring their hind flippers forward. They can thus only wriggle about rather clumsily on land. The eared seals (family Otariidae) have prominent ear flaps and they can bring their hind flippers forward to act as legs, thus allowing them to lollop over the ground at surprising speeds. This group includes the fur seals and the sea lions, the latter being distinguished by their relatively broad snouts. The walrus (family Odobenidae) can also bring its hind flippers forward, but it is easily distinguished from the eared seals by its massive ivory tusks.

Cold water fishermen

The majority of the 30 or so species of seals live in the polar and subpolar regions, and even those that live at lower latitudes are generally associated with cold currents. As we have seen with the penguins (page 180), the cold waters provide the richest fishing grounds. Fish and squid are the major foods of the seals, although they will eat whatever they can get. Starfish, crustaceans, molluscs, and even penguins and other seals figure in their diets but, as we shall see, a few species do have preferences.

The senses

It is generally agreed that the sense of smell is poorly developed in seals and that it plays little part in the detection of food. This is to be expected in an animal that has its nostrils closed while hunting, although it is possible that the rhinarium – the fleshy pad on the snout – may pick up the scent of fish shoals in the vicinity. This same organ may also be used by the females to find the right pups when they return to feed them on the beaches. The sense of taste is also poorly developed, and the seals rely to varying degrees on their sight, hearing, and sense of touch to find their food.

The eyes of a seal are remarkably large and well adapted for use both in and out of the water. They clearly play an important role in the animal's life. When underwater, where conditions are generally rather dim, the pupil opens almost to the full width of the eyeball and lets in the maximum amount of light. There is a tapetum behind the retina as well (see page 69), and the seal can obviously see quite well even in very murky water.

Among land vertebrates, most of the focusing is done when the light rays bend on passing from the air into the cornea at the front of the eye. This does not happen in water, because the cornea is not much denser than the water and the rays bend very little. The seals, like the fishes and the whales, focus with the aid of an almost spherical lens. This is fine for underwater use, but everything goes out of focus when the animal leaves the water. The seals have solved this problem with the aid of a special cornea and a pupil that contracts to a vertical slit in bright light. The cornea is irregularly curved and focuses light more efficiently in the vertical plane than in the horizontal plane. Light rays passing through the cornea and on through the vertical pupil thus form a clear image on the retina at the back of the eye.

Vision is blurred on land at night, when the pupil opens up, but this does not matter because the seals normally go back into the water to hunt at night, and those that stay ashore go to sleep.

The eyes obviously play a role in food-finding, but they are not the most important sense organs, as shown by the several blind seals that have been caught in well-fed condition. Polar seals also survive and feed without problems during the winter, when there are prolonged periods of complete darkness.

Hearing is probably much more important than sight as far as feeding is concerned, and there is good evidence that sea lions and several other seals use a system of echolocation (see page 160) to detect fishes and other prey swimming in the neighbourhood. Microphones hung close to seals in the water certainly pick up a lot of high-frequency sounds, and it is possible that all seals use some form of echolocation.

The moustachial whiskers, or vibrissae, also play a significant role in food-finding. Their bases are very well supplied with nerves and they are extremely sensitive to touch. They can detect currents set up by passing animals, and thus alert the seal to the presence of potential prey, and they can also help distinguish between edible and inedible objects when they touch them. In this respect, the vibrissae are of particular

The leopard seal— a savage loner
Left A solitary leopard seal conceals itself under the overhang of an ice floe and waits for an emperor penguin to dive into the water.

Left The seal shoots out from its hiding place to grab the unwary victim. This strategy decreases the chance of the penguin escaping by 'flying' through the water at speeds up to 55kph (34mph).

Below The leopard seal seizes its victim by the rump as the other penguins flee for their lives and leap in spectacular fashion back to the safety of the ice floe.

The waiting game
Above A leopard seal patrols close to the coast, waiting for an Adelie penguin—one of its favourite foods – to dive into the water.

value to bottom-feeding seals, such as the walrus and the bearded seal.

The chase

Having detected food in the vicinity, the typical seal or sea lion simply gives chase. There is no possibility of stalking the prey, although the seal generally approaches from below, from which direction it is least likely to be detected. The true seals swim like fishes, with the hind flippers held together to form a kind of tail fin that is moved powerfully from side to side. The front flippers are held close to the body except when the seals change direction. The eared seals obtain their forward motion by using their front flippers as oars, with the hind flippers being trailed close together as a rudder.

Most seals press home the chase only when the prey is of a size that can be swallowed whole. The victim is simply grabbed in the jaws, manipulated into the right position, and swallowed head first – thus ensuring that spiny fins do not scratch or stick in the throat. Larger prey is sometimes caught, however, and brought to the surface to be torn to

pieces by the seal's teeth. The latter are basically pointed pegs of varying lengths, designed for grabbing and restraining slippery prey; they can tear flesh, but they cannot chew. Little is known about the amounts of food eaten by the seals, but one estimate suggests that an adult grey seal cow, which may weigh anything from 60 to 100kg (132–220lb), eats about 7kg (15lb) of fish and other food per day.

Seal pups grow rapidly on the rich milk provided by their mothers, but they are not brought any other food and they are not taught to hunt. Many of them get their first solid meal by nosing around for crustaceans on the beach, and they often concentrate on crabs and shrimps during their first few weeks in the water, for these crustaceans are generally easier to catch than fishes.

The specialists

Although most seals will eat whatever they can catch, some specialize to a certain extent. The most specialized of the true seals is the crabeater seal (*Lobodon carcinophagus*) of the Antarctic. Its name is unfortunate, because it does not really eat crabs; it feeds mainly on the abundant planktonic crustaceans known as krill. Its teeth each have five prominent cusps, and those of the upper and lower jaws interlock loosely. When the animal has taken a mouthful of plankton it closes its jaws and forces the water out through the gaps between its teeth. The crustaceans trapped in its mouth are then swallowed. The ringed seal (*Pusa hispida*) feeds mainly on plankton around the Arctic coasts, but its teeth are less modified than those of the crabeater. It probably catches the larger crustaceans one by one.

The leopard seal (*Hydrurga leptonyx*) reaches lengths of about 3m (10ft) and is the largest of the Antarctic seals. Penguins and other seals figure largely in its diet, although it seems likely that only certain populations of leopard seals feed regularly on penguins. Small penguins are often swallowed whole, and the leopard seal has a specially con-

structed windpipe which enables it to do this. A normal windpipe is supported by a series of cartilaginous rings, but only the lower parts of these rings are present in the leopard seal, allowing the windpipe to be compressed without harm as bulky food passes along the gullet.

Leopard seals occasionally take penguins on the shore, but this is not common because the penguins can move much more quickly than the seal on land. Sea lions, on the other hand, can move more quickly than penguins, and the Australian sea lion (*Neophoca cinerea*) regularly catches penguins on beaches.

The walrus (*Odobenus rosmarus*) is the most aberrant member of the seal group – the Pinnipedia – in both looks and habits. It feeds mainly on clams and other molluscs, which it digs from the sea bed with its tusks. Its eyes are small and it detects the shellfish among the excavated mud by means of its sensitive vibrissae. The cheek teeth of the walrus are flat and usually very worn, from which it has long been assumed that they are used to crush the molluscs. Shell fragments are rarely found in the walrus stomach, however, and the fleshy parts of the prey seem to be swallowed whole. This suggests that the teeth merely crack the shells and that the walrus then sucks out the fleshy parts and spits out the shells. Such behaviour has recently been witnessed in captive animals. Other food includes starfish and various fishes, and the walrus may attack other seals and even small whales. Eskimos sometimes catch walruses by baiting strong lines with whale blubber, suggesting that the animals readily take carrion.

The bearded seal (*Erignathus barbatus*) from the Arctic Ocean is another bottom-feeder that depends largely on molluscs, but it does not compete with the walrus because it takes just those molluscs resting on the seabed. Like the walrus, it detects its food by means of its whiskers, and it probably feeds by cracking the shells and sucking out their contents in the same way. Its teeth are worn, but shell fragments are rarely found in the stomach of this rather heavy-bodied seal.

Above The seal bites its victim several times and shakes it vigorously until it is dead. Sometimes the head and legs are bitten off.

Right The seal tears at its victim and continues to thrash it about to loosen the skin, which may peel right away from the body and slip up round the neck.

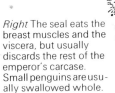

Right The seal eats the breast muscles and the viscera, but usually discards the rest of the emperor's carcase. Small penguins are usually swallowed whole.

Whales-Hunters Returning to the Sea

There are two main groups of whales: the large, toothless baleen whales, which filterfeed on small shrimp-like crustaceans called krill; and the smaller, toothed whales, which actively hunt larger prey. This fascinating group of mammals have returned to the sea and adopted a fish-like form, spending their entire lives playing, hunting, mating and even giving birth in the oceans.

Their smooth, torpedo-shaped bodies are devoid of external appendages apart from the fins. The hind limbs have disappeared, and the arms have become paddle-like 'fins.' The tail is flattened and expanded into large fibrous flukes, without bones, and some forms have a strong fibrous dorsal fin for extra stabilization when swimming and manoeuvring at speed. To reduce drag through the water whales, dolphins and porpoises have lost their body hairs, except for a few sensory bristles around the head and mouth in some species. The skin is usually smooth and slippery.

Despite their fish-like appearance, these amazing creatures swim not with a sinuous side-to-side movement, but by flexing their spine and moving the tail in an up and down motion. The large, flattened flukes produce the forward thrust. The dorsal position of their air-filled lungs keeps them the correct way up, aided by the dorsal fin in some species. The twisting and turning movements performed at speed are controlled by alterations in the positions of the 'fins' and flukes.

Seeing with sound

The sense of smell is virtually useless in whales, dolphins and porpoises (the cetaceans). Their sense of touch is only useful at very close range, and their eyesight, whilst usually moderate, is limited underwater over long distances or in dark depths. Hearing has become the primary sense of the cetaceans, the ears being specially modified for hearing underwater. They are insulated from the skull, which can transmit interfering sounds, by spongy tissue. This means that sounds reach the two ears independently so that the direction of the source can be located. As the displacement movements in vibrations in water are only one sixtieth that of air, a much heavier and more rigid transmission system is required in the middle ear. This is also essential for the reception of the higher frequencies used by the whales.

Whales emit sound waves, which they bounce off objects in their surroundings and receive as echoes, very like the sonar systems used to measure the depth of the seabed or to locate submarines. For general information whales send out low frequency clicks, which have great penetration but little resolving power. When communicating between themselves they use higher frequency clicks and whistles. The highest frequency sounds are used when hunting food, and whales can recognize their quarry by the way these vibrations are reflected and absorbed. Movements of prey must also play some part in recognition, although the bottle-nosed dolphin can distinguish between a piece of dead fish and an inanimate object of the same shape.

The giant filter feeders

The baleen whales have an arched upper jaw to accommodate the long keratin plates. The outer edge of each baleen plate is smooth and straight while the inner edge has a hairy fringe to trap the food; it continues to grow throughout the life of the whale. Although their hunting strategies are not so exciting as those of the toothed whales, these leisurely filter-feeding giants kill thousands of creatures at a time. Their main prey is the surface shoaling krill, but shoals of small fish such as sardines, herrings and anchovies are also swallowed. These whales live in small groups that migrate to warmer waters for breeding and to cooler polar waters for food. Their blubber is much thicker than necessary for bodily insulation so it has been suggested that it forms a food store to tide them over the lean periods during their long seasonal migrations.

The baleen whales employ two filtering methods. The slow-swimming right whales, with their high, arched jaws, swim through the krill shoals with their mouths open. The whale's forward movement causes a build up of pressure, which forces the water out between the

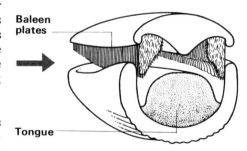

Baleen plates

Tongue

The filtering apparatus of the blue whale
Above The mouth opens and reveals the great curtains of whalebone, or baleen. Vast amounts of water and krill are taken into the mouth.   *Below* The mouth closes and water is forced out through the baleen. Krill is held back and swept into the throat by the tongue.

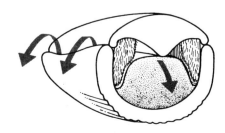

Intelligent hunter
Below The killer whale lifts its 'smiling' face from the water and reveals the sharp, conical teeth that it uses to maintain a secure grip on its prey. The teeth can tear flesh, but most prey is swallowed whole.

baleen plates. The krill are trapped on the hairs of the plates and at intervals the mouth is closed and the large grooved tongue collects the food and pushes it to the back of the mouth, where it is swallowed. The mouth is usually closed during a dive.

The gigantic blue whale, the largest living animal, and its fast swimming relations such as the rorquals, which have shorter baleen plates, swim into the shoals, open their mouths and suck in thousands of prey with each mouthful of seawater. The mouth is closed again and the water immediately forced out between the baleen plates. The tongue then sweeps the food from the baleen plates, takes it to the back of the mouth, and pushes it down the throat. The mouth then opens again for another gulp.

Deep-water hunters

The largest of the toothed whales is the sperm whale, made famous by the fictional whale, Moby Dick. These whales, with their huge, bulbous foreheads, use sonar to hunt for squid and octopus in the ocean depths. They migrate for feeding and breeding purposes from the poles to the tropics.

Killer whales in a co-ordinated attack
Above left Killer whales hunt in packs of up to 60 individuals. Here, a small pack has discovered a crabeater seal asleep on an ice-floe and they are surrounding the ice.

Left One whale rears out of the water and uses its weight – up to 8 tonnes when adult – to tilt the ice. Other whales may push up from the other side. The seal slides off the ice and is rapidly caught by one of the patrolling killers.

Left The whale bites and crushes the prey with its powerful teeth. Small prey, such as fish and penguins, are swallowed whole, but larger prey is torn to pieces by repeated biting and shaking, with several whales taking their share.

The teeth of these whales are unlike those of the other mammals, being conical and peg-like with a single root and a pulp cavity that degenerates when the tooth stops growing. They do not erupt until the whales are about nine or ten years old. They are not used to chew the food, but for fighting. Squid and fish recovered from a sperm whale's stomach rarely show teeth marks but are swallowed whole, usually undamaged. The large, aggressive males take deep water squid between 1·3 and 2·2m (4·3–7·2ft) long but the stomach of one enormous bull sperm whale contained a squid of 10·49m (34·4ft) in total length. The scars of suckers measuring 10cm (4in) have been found on bull sperm whales. Scars caused by the cephalopods are rarer on the females, suggesting that they limit themselves to the smaller squid and octopus, which do not struggle very much.

The fish found in sperm whale stomachs are deep water species such as shark (one 3m (10ft) blue shark being recovered in one case), cod, snapper, skate, angler fish and rattails. Sperm whales have long throat grooves and these probably allow for distention of the gullet when large prey is swallowed.

The beluga
This small, white, domed-head whale lives in the Arctic, feeding on arctic char, flounder, halibut, capelin, squids, crustaceans and other bottom-dwelling creatures. It is known as the sea canary because of its loud liquid trill, which is readily heard above the waves even though produced in the ocean depths.

Killer whales
These intelligent whales, with their distinctive tall dorsal fin, hunt in packs of six to sixty individuals, taking any living creature that crosses their path, including other whales, dolphins and porpoises. Although an adult male killer measures only just over 9m (29·5ft), the pack will take much larger whales, such as the 20m (65·5ft) filter-feeding arctic right whale. Their usual food is other cetaceans, seals and sea lions, and penguins, but they will swallow sharks and squids. They have a very large gape and about twelve large, sharp, recurved, conical teeth that firmly interlock when the jaws close. The prey is usually swallowed whole; the remains of 30 seals have been found in the stomach of one killer whale, while another revealed 13 porpoises and

14 seals. When the killer whale bites a large animal it tears off a chunk of flesh by shaking its head to and fro.

They co-operate when hunting, encircling their victims or cornering them in an area of sea enclosed on three sides by land. 'Sentries' are posted at the outlet to prevent escape whilst the other pack members attack in concert. They have been known to dive deeply and rush upwards, surfacing through ice up to 1m (39in) thick to dislodge unwary prey such as seals and penguins. When a floe is too thick to smash, killers are reputed to surface and lean on the ice in an attempt to rock it and slide sleepy seals into the jaws of the encircling pack.

Playful porpoises and dolphins
These smaller cetaceans also hunt in packs of several individuals, usually swallowing surface-shoaling fish and squid. The blunt-nosed porpoises and the beaked-nosed dolphins can torpedo through the water at speeds in excess of 46kph (29mph) and often shoot out in graceful leaps, even snatching flying fish from the air as they skim the waves. A few species have even entered estuaries and rivers in search of food.

The Polar Bear – King of the Arctic

The polar bear is an animal of the High Arctic. It is found mainly around the shores of the Arctic Ocean, but it is patchily distributed. The main populations are centred in Greenland, in the islands of Arctic Canada, and around the northern coast of Alaska. Smaller populations occur in the Laptev Sea area to the north of Siberia and also in the region between Novaya Zemlya, Franz Josef Land, and Spitzbergen. Polar bears also frequent much of the Hudson Bay region of Canada, and some individuals stray as far south as Newfoundland. They have been seen swimming in the sea more than 300km (186 miles) from the coast, and they are known to roam as much as 150km (93 miles) into the coniferous forests, but their true habitat is the pack ice, for this is where they find the seals that make up about 90 percent of their diet.

Stalking or ambushing

Several seal species help to fill the polar bear's stomach, but the most important is the ringed seal. Some seals are caught in the water, but the seals are generally much quicker than the bears in this element and the bears actually catch most of their prey on the land or on the ice. Harp seals and hooded seals, for example, are usually attacked on their breeding grounds, where the polar bears can take the helpless pups with great ease. And, using their sense of small, polar bears can detect young ringed seals in their burrows more than 2m (6·5ft) below the snow surface.

Adult seals resting on solid pack ice or frozen shores may be stalked by the bears, which rely mainly on their hearing and their incredibly keen sense of smell to detect the prey from a distance. Eyesight comes into play as the polar bears get closer to their prey, and the animals use any available rocks or blocks of ice to conceal their approach until they are within about 30m (98ft) of the seals. The bears then race over the ground to make the kill. The seals have no chance, for the roles are reversed on the land; the seal is slow and clumsy and the polar

bear, despite its lumbering appearance, can run smoothly at about 30kph (19mph). Its hairy soles prevent it from slipping on the ice.

Seals resting on small ice-floes have to be approached by water. The bear swims towards them with just its nose above water, doing the 'dog paddle' with its front legs and trailing its hind legs as a rudder. It dives as it nears the ice, and re-surfaces right by the edge. If a seal is near enough to the edge, the bear may knock it into the water with one blow of its enormous paw, but otherwise the bear leaps onto the ice and pounces before the seal has time to slip into the water and swim away.

The studies of Ian Stirling of the Canadian Wildlife Service have shown that on the extensive winter ice sheets only about one quarter of the polar bear's meals are obtained by hunting. The rest are obtained in a much less energetic way by using the 'blow-hole technique'. The bear takes up station by a seal's breathing hole and sits very still, perhaps for hours, until a seal comes to that particular hole for air. That is the end of the seal, for its skull is crushed by one swipe of the bear's paw — the bear's standard method of despatching its prey. The blow may well lift the seal right out onto the ice, but if not, the bear has no difficulty in dragging the dead seal out with its jaws.

A really hungry polar bear eats just about every part of a captured seal, but in times of plenty the bears are wasteful and they probably kill far more seals than they need to eat. Very often just the skin and blubber are stripped from the seals, especially the young ones, and the rest of the carcase is left for the scavenging ravens and the arctic foxes.

The other ten percent

Seals may contribute about 90 percent of the diet of the ordinary polar bear, but the remaining 10 percent may be obtained from a wide variety of sources, both on land and in the water. Young musk oxen are sometimes killed, and the bears also catch a certain amount of fish.

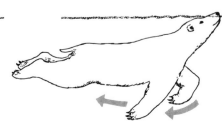

An easy meal for the polar bear
Above The polar bear smells a ringed seal pup in its birth chamber under the snow and ice. It begins to sweep the snow from above the chamber.

Seaweed is enjoyed on occasion, and during the summer moult, when the bears remain on land, they eat quite a lot of grass and berries.

The bears living in the Hudson Bay region regularly eat vegetable matter, spiced with rodents and other small mammals, but geese and ducks account for about 70 percent of the polar bear's diet in this region. Some are taken at their nests, and the eggs are also eaten, but most of the birds are caught on the water. The bears dive under the floating flocks and come up to snatch them from below. Relatively few birds are eaten in other parts of the bear's range.

Cannibalism occurs from time to time, but does not play a significant part in the diet. Carrion in the form of a stranded whale provides a real feast, and the normally solitary polar bears may then gather in quite large numbers to carve up the carcase. Overall, the polar bear

Duck-hunting among the ice-floes
Below A polar bear notices a flock of long-tailed ducks sleeping on the water some distance away. Moving backwards, it slides silently into the water. Turning under the water, the bear swims strongly towards the birds with just its snout above the surface. It uses its powerful front legs and broad paws as paddles, and trails its hind legs as a rudder. It comes up very close to the sleeping ducks and makes a sudden

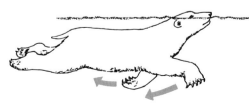

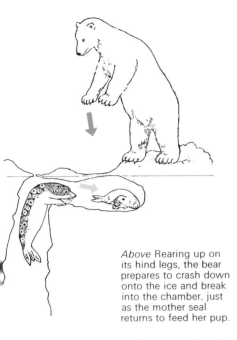

Above Rearing up on its hind legs, the bear prepares to crash down onto the ice and break into the chamber, just as the mother seal returns to feed her pup.

Above The mother ringed seal retreats rapidly as the bear breaks into the chamber. The cub is killed with a single blow from the bear's great paw.

Above The baby seal is hooked out of the ruins of its nursery by the bear's paw and grabbed by the enormous jaws. Only the skin and blubber may be eaten.

is the most carnivorous bear species; the others – the brown bears and the various kinds of black bears (see page 106) – are primarily vegetarian or omnivorous.

Do they use tools?

Many Eskimo stories and carvings suggest that the polar bear uses rocks or blocks of ice to kill seals. The use of such 'tools' seems unnecessary for an animal that can crush a seal's skull with one swipe of its paw, but scattered observations indicate that they may be used on occasion – possibly to break into the frozen den of a seal or to bring down a seal that is just out of reach. The polar bear can certainly throw sizeable objects with considerable force, for this has been seen many times in zoos. Scientists with the Canadian Wildlife Service have also discovered that some polar bears are able to spring the traps that they set by dropping small rocks onto them.

lunge at one of them, grabbing it securely with the immense canine teeth and crushing it to death instantly. This form of duck-hunting is confined largely to the Hudson Bay area, where birds are a major prey.

Savage majesty
Right The savagely majestic polar bear (*Thalarctos maritimus*) reaches weights in excess of 500kg (1100lb), the average weight of a male being 410kg (902lb). Females are usually a little smaller.

Man the Hunter

Less keen of hearing, and with poorer sight and sense
of smell than most hunters, man also lacks speed and strength; but he has
some much more valuable assets — his hands and his brain have enabled
him to fashion tools and weapons to increase his power and
to outwit his prey. He now has the power to destroy all life on earth.

Man the Hunter

'You are looking at the most dangerous animal in the world. It alone of all the animals that ever lived can exterminate (and has) entire species of animals. Now it has achieved the power to wipe out all life on earth.'

The above words appear under an exhibit in New York's Bronx Zoo, and the exhibit is — a mirror!

The human animal is known scientifically as *Homo sapiens* because of his superior brain power or intelligence. This factor has enabled man to rise above the other animals and to dominate them in a way never known before. Man has become not just a top predator, but a super-predator, sitting at the top of a vast number of food chains, both on land and in the sea, with the power to manipulate them for his own requirements. The ability to exploit and manipulate so many food chains has allowed man to spread all over the world, with the result that *Homo sapiens* has become one of the commonest species on earth. And this is precisely why mankind is such a dangerous creature.

Simple beginnings

When man first walked the earth— at a date continually being pushed back by the anthropologists and currently standing at something over three million years ago — he was ecologically no different from other omnivorous creatures. He led a simple hunting and gathering existence in a fairly well-defined home range, catching what animals he could and exploiting the various plant foods as they came into season. His numbers were kept in check by food supplies, and he was therefore just another component of the intricate web of life, causing only temporary changes in his surroundings.

Throughout his history, man has always remained unspecialized in physical terms; he has no built-in weapons such as powerful teeth or claws, and he has not even any real running ability. But his upright stance freed his hands to use extraneous weapons or tools, and this was a very great advance, especially when considered in conjunction with increased brain power. Tools can be picked up and discarded quickly, according to the situation, and different tools can be used for different purposes.

Man's earliest weapons were undoubtedly stones, which were picked up and thrown at animals in attempts to kill them. Such attempts were quite successful, judged by the many damaged baboon skulls found in various parts of Africa in association with rounded stones and the remains of early men. The limb bones of animals were probably used as clubs from very early times, and broken

branches must have been used in the same way. Branches would also have been used as spears, and gradually made more effective as men learned how to sharpen them, either with fire or with stone tools. Bones were also sharpened and used as daggers or spears. Stone tips were later fixed to the wooden spears, and wooden spear-throwers were also developed, enabling the hunters to increase the range of their weapons significantly by effectively increasing the length of the throwing arm.

By Neanderthal times, some 120,000 years ago, men had become very successful hunters of big game in Africa and many other parts of the world. They worked in groups like hunting dogs, but their intelligence allowed them to plan their hunting strategies in advance. They could drive prey into pre-planned ambushes, for example, and Neanderthal Man himself probably employed large pitfall traps to capture the huge mammoths and woolly rhinos that shared the land with him.

During later Palaeolithic times and the succeeding Mesolithic Age, weapons and other tools were greatly improved as men became more skilled at working stone and bone. The bow and arrow was also invented, although we do not know just where or when, and it rapidly spread to most cultures. It allowed the hunters to kill at a much greater distance than they could with their spears. The dog was also domesticated a little over 10,000 years ago as a useful hunting aid as well

Australian fish-trap
Left Aborigines from Australia's Arnhemland remove a small shark from a drum-net fish trap at low tide. The fishes enter the trap when the tide is in, but they cannot escape as the tide falls.

Plenty of puff
Right This blow pipe being used in the forests of Borneo looks very unwieldly, but its length gives the dart increased speed, and the natives become so proficient in the use of such pipes that they can hit small birds sitting high in the trees. Some tribes poison the tips of their darts to increase their effective killing power.

Two ways of increasing the range
Left Energy stored in the drawn bow gives the arrow much greater speed, and therefore greater range, than it can receive from the Bushman's arm alone.
Right The Australian Aborigine effectively increases the length of his arm, and therefore the range of his spear, by holding the spear in a simple spear-thrower.

as a companion. It was used — and still is in one form or another — to detect, catch, or retrieve prey. Cultures based on fishing grew up in areas with rich fish supplies and became more sedentary than the other hunting and gathering peoples. A range of very efficient bone hooks and harpoons was developed by the various fishing communities, and nets were woven from bark and other plant fibres.

Despite the great advances in weaponry, the Mesolithic people remained at the mercy of their surroundings. They could hunt and gather only what the environment provided, and their numbers were still limited by food supplies. A big step forward came when men started to herd their own animals instead of hunting them. Reindeer, camels, goats, and several other hoofed mammals were all being herded in various parts of the world more than 9,000 years ago. But the real turning point came with the birth of agriculture some 8,500 years ago, when men learned how to grow their own crops. Populations settled down around their farms and, freed from the need to look for food every day, the people began to put their brains to work in other directions. With a secure and controllable source of food, and with a wide range of weapons and tools to help them, the people gradually freed themselves from most of the environmental controls to which they had previously been subject. Numbers began to rise, and so emerged the world's most dangerous animal — modern man.

Primitive hunters today

Agriculture originated in different parts of the world at different times, but the three main centres were Western Asia, China, and Central America. The practice spread out from these regions into nearly every part of the world, and almost every community today depends upon some form of crop-growing. There are, however, a few cultures which, even today, rely on the ancient hunting and gathering system.

Today's hunting and gathering people probably number no more than about 250,000, distributed rather thinly across the globe in areas where agriculture made no impact during its initial spread. Some live in tropical forests, where food is plentiful all the year round and there

was no incentive for the people to abandon their traditional way of life for agricultural pursuits. Other hunters live in the cold northern regions and in the deserts — both areas where agriculture is not possible without advanced technology. A fourth group of present-day hunters and gatherers is represented by various fishing communities. Very few of these primitive cultures are entirely unaffected by modern life — the Eskimos use rifles, for example, and most of the fishing communities use modern lines and nets — but they still maintain their fundamental ties with the environment.

One of the most primitive human cultures still in existence, albeit a very fragile existence, is that of the Australian Aborigines. These people arrived in

Australia with a Stone Age culture between 20 and 30 thousand years ago, and they changed very little until the Europeans arrived. It is believed that there were about 300,000 Aborigines in Australia when the European settlers moved in 200 years ago, and that this was close to the maximum possible number that the primitive hunting and food-gathering techniques could support in what is a largely hostile land.

Less than 40,000 pure-blood Aborigines survive today, and probably no more than 1,000 of these still follow their traditional way of life. They live in small groups and wander over their home ranges with very few possessions – some simple spears and boomerangs, some flint knives, a number of digging sticks used for unearthing plant roots and tubers, a few wooden dishes, and string bags made from plant fibres. They have no bows and arrows, although they do use spear-throwers. They also use fire, and the women try to carry smouldering wood with them on each day's march so that they can easily start a fire at the new camp site in the evening. The fire is used for cooking, and it is also important for keeping warm at night in the desert.

During the daily walk, the women search for plant food. They have a wonderful ability to detect food-filled tubers in the ground by recognizing the shrivelled remains of the aerial parts of the plants, and they are also skilled at extracting the famous wichiti grubs from tree stumps and roots and at finding the nutritious honeypot ants under the ground. Hunting is the men's job. They fan out on each side of the group and hope to flush out wallabies, lizards, and other animals that they can strike down with their spears. The area covered depends on the season and also on the nature of the terrain. In the most favourable areas a family group can obtain enough food without walking more than a few kilometres, and may then keep to one camp site for several days or even weeks. Coastal Aborigines, who can rely on plenty of fish to eat, stay put for months. But those groups living in the arid interior of the continent have to travel many kilometres every day in order to find enough food. They come to know every part of their range in minute detail – where they are likely to find water, where they might find honey ants, and so on.

Other present-day people with primitive hunting and gathering economies include the Eskimos, the Bushmen of the Kalahari, the Semang of Malaysia, the Pygmies of Central Africa, and various groups of Indians in South and Central America. These all have the bow and arrow, and most of the jungle tribes

Seal slaughter – an annual controversy
Above Fur seals are brutally clubbed to death in the Pribilof Islands. The annual culling of seals may sometimes be necessary to protect fisheries, but wholesale slaughter of seals for skins always leads to arguments between hunters and conservationists.

improve its efficiency by poisoning the arrow tips with toxic extracts from plants or animals, such as arrow-poison frogs.

Man the killer
Today's hunters and gatherers are on the same ecological plane as the earliest men, in equilibrium with their surroundings and causing no long-term changes in it. The same cannot be said of modern man, who has freed himself from most of the natural ecological controls by means of agriculture and technology. He no longer has to hunt animals for food, but he remains a killer and is having a devastating effect on the world's wildlife.

Only in the waters does man still practise extensive hunting for food, but even the once-bountiful seas are beginning to suffer from overfishing. Bigger and more powerful boats are hauling far greater quantities of fish from the water than were ever taken in the past, and if this goes on for much longer there will be few fish left to catch. This has already happened with the larger whales.

On land, man kills on a large scale to protect his property. This usually means

Harvest of the sea
Left Millions of fishes and other marine animals, such as crabs and squid, are dragged from the sea every day. These baskets contain just a fraction of one boat's daily haul from the Gulf of Thailand. Without controls or quotas, there will soon be no fish left for the fishermen to catch.

The inedible quarry
Right The red fox is one of the most widely hunted of land animals, although its flesh is useless and its fur almost so. It is hunted for 'sport' and also because it kills a number of game birds and poultry, but it must be admitted that hunting has had little effect on the fox's population.

killing predatory animals that would otherwise kill his domestic stock, but it also involves killing herbivores, such as kangaroos, that compete with domestic stock for grazing. Commercial killing, mainly to obtain animal skins, is another form of slaughter taking place on a large scale. Some animals have always been killed for their skins, but the killing is now big business and many species have declined rapidly as a result. The most familiar examples are the large cats, whose beautiful coats have long been in demand for fashion. Many countries now have laws to prohibit the killing of these animals and the export or import of their skins, but the killing still goes on.

Predators that once stood securely at the top of their food chains are now preyed on mercilessly by man — and needlessly as well, for no-one needs coats made from wild animal skins in these days of man-made fabrics.

Most reprehensible of all is the killing of wild animals for so-called sport, or simply because they are there. At one time the competition might have been more or less equal, but now that man has powerful rifles and other technological equipment animals have little chance of escape. Many are becoming exceedingly rare, and some have already disappeared for good. The passenger pigeon, for example, probably numbered in excess of 5,000 million birds in North America early in the 19th century, but guns blasted it from the skies and nests were plundered for their fat, nutritious chicks; the last bird died in a zoo in 1914. A similar fate almost befell the majestic bison, and could await desert-living gazelles, which are shot from fast-moving vehicles, and polar bears, which are hunted by helicopter. It is true that such destruction could not take place without today's sophisticated equipment, but we must not forget that a gun cannot kill by itself: the killer is the man behind it.

The direct killing described above has been responsible for the extinction of at least 50 species since the dodo in 1680 and the decline of very many others, but indirect killing through the destruction of wild habitats is an even more serious threat to wildlife. Vast tracts of natural vegetation are being destroyed to make way for agriculture and industry, com-

Once threatened
Above The American bison was almost shot out of existence in the 19th century, but conservation measures saved it just in time.

Gone forever
Below The dodo of Mauritius had no enemy until man arrived on the island. Within 200 years every bird had been destroyed.

munications, and houses. Allied to this is the loss of wildlife through pollution of the air and land with industrial wastes and pesticides. This can all be blamed on man's increasing population, and it is a vicious spiral: the more land we cultivate, the more people it can support; and these people themselves generate a further increase in population that demands more land. The only way out if we are to retain any semblance of a natural world around us is to stabilize or, better still, reduce the human population. This should not be beyond man's ability.

Perhaps one day a zoo will be able to exhibit a mirror with the label 'You are looking at the world's most intelligent species. It has the power to destroy all life, but it prefers to manage the environment so that the greatest possible number of species can exist in a state of equilibrium.' This day has not yet arrived.

A tarsier, virtually the only wholly carnivorous primate, munches a cicada by night.

Further Reading

General

Angel, M. & Harris, T. *Animals of the Oceans* Peter Lowe 1977
Burton, M. *The Sixth Sense of Animals* Dent 1973
Burton, M. & Burton, R. *Encyclopedia of Amphibians, Reptiles, and other Cold-Blooded Animals* Octopus 1975
Burton, R. *The Senses of Animals* David & Charles
Cohen, I. E. *The Predators* New Burlington Books 1978
Cott, H. B. *Looking at Animals* Collins 1975
Dozier, T. A. *Dangerous Sea Creatures* Time Life Television 1976
Fogden, M. & Fogden, P. *Animals and their Colours* Peter Lowe 1974
Hardy, Sir A. *The Open Sea: The World of Plankton* Collins 1956
Hardy, Sir A. *The Open Sea: Fish and Fisheries* Collins 1959
Parks, P. *The World You Never See – Underwater Life* Hamlyn 1976
Reader, J. Croze, H. *Pyramids of Life* Collins 1977
Readers Digest, *The Living World of Animals* 1970
Street, P. *Animal Weapons* MacGibbon Key 1971
Whitfield, P. *The Hunters* Hamlyn 1978

Invertebrates

Blaney, W. M. *How Insects Live* Elsevier-Phaidon 1976
Bristowe, W. S. *The World of Spiders* Colins 1958
Burton, M. & Burton, R. *Encyclopedia of Insects and Arachnids* Octopus 1975
Corbet, P. S. *A Biology of Dragonflies* Witherby 1962
Crompton, J. *The Hunting Wasp* Collins 1948
Crompton, J. *The Spider* Collins 1950
Evans, H. E. & Eberhard, M. J. W. *The Wasps* University of Michigan 1970
McKeown, K. C. *Australian Spiders* Angus & Robertson 1963
Main, B. Y. *Spiders* Collins Australian Naturalist Library 1976
Mash, K. *How Invertebrates Live* Elsevier-Phaidon 1975
Morton, J. E. *Molluscs* Hutchinson 1967
Nichols, D. *Echinoderms* Hutchinson 1969
Oldroyd, H. *The Natural History of Flies* Weidenfeld & Nicolson 1964
Savory, T. H. *The Spider's Web* Warne 1952
Schneirla, T. C. *Army Ants* Freeman 1971
Smith, E. et al *The Invertebrate Panorama* Weidenfeld & Nicolson 1971
Yonge, C. M. & Thompson, T. E. *Living Marine Molluscs* Collins 1976

Mammals

Bertram, B. *Pride of Lions* Dent 1978
Bueler, L. E. *Wild Dogs of the World* Constable 1974
Burrows, R. *Wild Fox* David & Charles 1968
Burton, M. *How Mammals Live* Elsevier-Phaidon 1975
Burton, M. & Burton, R. *Encyclopedia of Mammals* Octopus 1975
Coffey, D. J. *The Encyclopedia of Sea Mammals* Hart Davis 1977
Dominis, J. & Edey, M. *The Cats of Africa* Time Life 1968
Eaton, R. L. *The Cheetah* van Nostrand Reinhold 1974
Ewer, R. F. *The Carnivores* Weidenfeld & Nicolson 1973
Fox, M. W. (Ed) *The Wild Canids* van Nostrand Reinhold 1975

Guggisberg, C. A. W. *Wild Cats of the World* David & Charles 1975
Harrison, R. J. & King, J. E. *Marine Mammals* Hutchinson 1965
Hinton, H. E. & Dunn, A. M. S. *Mongooses, their Natural History and Behaviour* Oliver & Boyd 1967
Larsen, T. *The World of the Polar Bear* Hamlyn 1978
Lawick, H. van *Savage Paradise* Collins 1977
Lawick-Goodall, H. & J. van *Innocent Killers* Collins 1970
Martin, R. M. *Mammals of the Seas* Batsford 1977
Mech, L. D. *The Wolf* American Museum of Natural History 1970
Neal, E. G. *The Badger* Collins 1948
Neal, E. G. *Badgers* Blandford 1977
Perry, R. *The World of the Tiger* Cassell 1964
Perry, R. *The World of the Polar Bear* Cassell 1966
Perry, R. *The World of the Walrus* Cassell 1967
Perry, R. *The World of the Jaguar* David & Charles 1970
Petersen, R. *Silently by Night* Longmans 1966
Schaller, G. B. *The Deer and the Tiger* University of Chicago Press 1967
Schaller, G. B. *The Serengeti Lion* University of Chicago Press 1972
Yalden, D. W. & Morris, P. A. *The Lives of Bats* David & Charles 1975

Fishes, Amphibians, and Reptiles

Bellairs, A. *The Life of Reptiles* Weidenfeld & Nicolson 1969
Budker, P. *The Life of Sharks* Weidenfeld & Nicolson 1971
Burton, M. & Burton, R. *Encyclopedia of Fish* Octopus 1975
Echternacht, A. C. *How Amphibians and Reptiles Live* Elsevier-Phaidon 1977
Guggisberg, C. A. W. *Crocodiles* David & Charles 1972
Marshal, N. B. *The Life of Fishes* Weidenfeld & Nicolson 1965
Oulahan, R. *Reptiles and Amphibians* Time Life Television 1976
Parker, H. W. & Grandison, A. G. C. *Snakes – A Natural History* British Museum (Natural History) 1977
Whitehead *How Fishes Live* Elsevier-Phaidon 1975

Birds

Baker, J. A. *The Peregrine* Collins 1967
Bradbury, W. *Birds of Sea, Shore, and Stream* Time Life Television 1976
Brown, L. *African Birds of Prey* Collins 1970
Brown, L. *British Birds of Prey* Collins 1976
Brown, L. *Birds of Prey* Hamlyn 1976
Brown, L. *Eagles of the World* David & Charles 1976
Burton, R. *How Birds Live* Elsevier-Phaidon 1975
Eastman, R. *The Kingfisher* Collins 1969
Everett, M. *Birds of Prey* Hamlyn 1975
Everett, M. *A Natural History of Owls* Hamlyn 1977
Sparks, J. & Soper, T. *Owls: Their Natural and Unnatural History* David & Charles 1978
Tubbs, C. R. *The Buzzard* David & Charles 1974
Watson, D. *The Hen Harrier* T. & A. D. Poyser 1977

A cannibalistic female mantis has begun to eat her mate even before copulation is over.

Index

Figures in italics refer to illustrations.

Credits

Picture Credits
The publishers wish to thank the following photographers and organizations who have supplied photographs for this book. Photographs have been credited by page number and position on the page: (B) Bottom, (T) Top, (BL) Bottom left, etc.

Photographs
Heather Angel: Copyright page, 38, 40(R), 43(T), 88(TL), 92(B), 196(T);

Photo Aquatics: 188 (P Kopp);

Aquila Photographics: 49(E K Thompson), 63(BR, D I McEwan), 134(R, D Richards), 146(Don Smith), 157(R, F V Blackburn), 159(A T Moffet), 184(B, W G McIlleron);

Ardea Photographics: 24(R, Wardene Weisser), 25(M D England), 14-15(E Mickleburgh), 92(T, McDougal), 114(Adrian Warren), 117(B, Hans Beste), 119(L, Hans/Judy Beste), 125(B, John Wightman), 198(M D England), 207(I R Beames);

Axel Poignant: Second contents spread (TR), 208-9, 210(B), 211(T);

S C Bisserot FRPS: 79(BL), 192(B), 196(BR);

N A Callow: 20(T), 79(TR), 83(CL), 149(B), 152(T);

Michael Chinery: 32, 33, 41(B), 54(BL), 64(TR), 69(TR), 81, 88(CR), 123(T), 153(B), 212(B), 216;

Gabrielle Churchouse: 56-7, 128;

Ms Densey Clyne: 34, 54(T), 55, 117(T), 214;

Bruce Coleman Ltd: Endpaper (Simon Trevor), Title page (Masud Qureshi), First contents page (Jane Burton), 18(Stouffer Productions), 20(BL, John Shaw), 21(Jeff Foott), 29(L, M P L Fogden), 50(L, John Shaw), 68(Jane Burton), 75(C, Jane Burton), 77(Mark Boulton), 85(T, David Hughes), 86(Jane Burton), 87(T, Jen and Des Bartlett), 90(John and Sue Brownlie), 97(L), 100(Joe Van Wormer), 101(Hans Reinhard), 102-3 (Jane Burton), 107(T, Jeff Foott, B, Jonathan Wright), 109(T, C, Mary Grant), 112(D Robinson), 113(N Myers), 115(T, Stouffer Productions), 118(T, Francisco Erize), 121(Des Bartlett), 127(B, N Myers), 130(Stouffer Productions), 131(Bob and Clara Calhoun), 132(Leonard Lee Rue), 137(E Breeze Jones), 156(Hans Reinhard), 157(L, Dennis Green), 165(Lee Lyon), 170(B, Fritz Polking), 173(Jane Burton), 183(TR, David Hughes), 184(C, David Hughes), 190(T, Hans Reinhard), 190(B, Jane Burton), 191(Hans Reinhard), 192(T, Jane Burton), 193(TL, Hans Reinhard), 194(B, Oxford Scientific Films), 197(R, R and M Borland), 199(Jen and Des Bartlett), 200(L, Hans Reinhard), 200(R, R and M Borland), 202-3(Francisco Erize), 213(BL, Leonard Lee Rue III);

Morris H Colthorpe: 70(C);

Gene Cox: 175(TL);

Gerald Cubitt: 105, 134(L);

Adrian Davies: 181(TR);

M D England: 110, 150(B);

Inigo Everson: 181(TL, BC);

Robin Fletcher: Second contents spread, 24(L), 30, 74, 95(T), 153(T);

Ron and Christine Foord: 51, 65;

Keith Gillett: 39(CR), 41(T), 42(R), 175(BL, BR),

Robert Harding Associates: 147(Brian Hawkes), 193(TR), 211(B);

D P Healey: 69(B), 73(C), 85(B), 87(B), 179(CL);

David Hosking: 96, 162(T);

Eric Hosking FRPS: 16, 97(R), 119(R), 142(B), 144, 150-1(T), 167(T), 170(T), 180, 182(T), 185;

Jacana: 66-7(Robert), 94(T, Fievet), 95(B, René Volot), 126(Robert);

Dick Jones: 50(TR), 53;

Dr Hans Kruuk: 23, 129;

Frank W Lane: 105(B, Arthur Christiansen), 111(JV Spalding), 125(T, Peter Davey), 140(Georg Quedens), 142(T, Peter Davey), 152(B, W T Davidson), 189(B);

Professor James E Lloyd: 75(T), 83(TR);

Michael Lyster, The Zoological Society of London: 59(B), 183(TL), 193(B);

John Mason: 36, 37, 46-7, 83(TL), 148(B), 161(TL), 178, 179(B);

Dr Pat Morris: 39(T), 42(TL), 52, 59(T), 61(B), 63(BL), 64(BL, BR), 71, 73(B), 89, 160, 161(BR), 162(B), 163, 166, 186, 201, 204, 212(T), 213(TR, CR);

Natural History Photographic Agency: Half-title (Stephen Dalton), 20(BR, Anthony Bannister), 44-5 (Stephen Dalton), 78(Stephen Dalton), 108(E Hanumantha Rao), 123(B, Anthony Bannister), 124(Peter Johnson), 155(Peter Johnson), 210(T, Peter Johnson);

Natural Science Photos: 29(R, P H Ward), 48(L, Curtis E Williams), 63(T, P H Ward), 73(T, P H Ward), 143(T, M Stanley Price);

Oxford Scientific Films: 35, 60, 69(TL), 99, 174(B), 194-5(T);

G R Roberts: 116, 177(BR);

Dr Edward S Ross: Author's Page, Foreword, 26-7, 31(TR), 70(T), 76, 84(CL, CR), 91, 93, 103(TR), 115(CL), 120, 122;

Royal Society for the Protection of Birds: 164(Sixten Jonsson), 169(M W Richards), 184(T, M W Richards);

John Topham Picture Library/Norman Myers: First contents spread(TL), 134-5(T);

B Tulloch: 151(R);

B S Turner: 75(B), 98, 138-9, 154, 158, 171, 182(B);

Michael Tweedie: 48(R), 80;

Dr Peter Ward: 47(B), 82, 84(T), 148(T), 149(T), 167(B);

Douglas P Wilson FRPS: 28, 39(CL), 40(L), 42(BL), 43(B), 174(T), 176(T), 176-7(B), 197(L);

J F Young: 141;

ZEFA: 109(B, Werner H Muller);

Artists
Copyright of the drawings on the pages following the artists' names is the property of Salamander Books Ltd.

Richard Blake: Author's page, 22(B), 139, 141, 156, 157, 164, 166, 168, 171, 181, 183, 185;

Peter Hayman: 145, 146;

Sally Launder: 115(B), 155;

Michael Lynn: Copyright page (L), Second contents spread (TL), 31, 47, 65, 102, 103, 110-11, 154, 198, 199(T), 202-3;

Lydia Malim: Copyright page (TR), First contents spread(TL), 16(T), 18, 19, 22(T), 40, 68, 70, 71, 72, 74, 90-91, 93, 116, 118, 144, 161, 176(B), 196, 197, 199(CL), 204;

Josephine Martin: 100;

Sean Milne: 134-5;

Alan Pearson: 143;

Gordon Riley: First contents spread(TR), 16, 32, 34, 37, 48-9, 50, 51, 52, 53, 54, 55, 56, 79, 81, 83, 86, 87, 179;

Barbara Tenison Smith: 158;

Christine Smith: 16(C), 38;

Juliet Stanwell Smith: 106, 115(T), 176(C), 201, 206-7;

Robina Smith: 98-9, 101;

Anthony Swift: 46, 186-7, 205;

Tyler/Camoccio Design Consultants: 58, 84, 89, 97, 104, 127, 128, 133, 190, 191, 195;

Acknowledgements
The publishers would like to thank W S Bristowe and Williams Collins and Sons for allowing the drawings on pages 32, 37, 50, 52, 53, 55(B) and 81 to be based on those in *The World of Spiders*. Thanks are due also to T H Savory and Dr Ian Wallace, Keeper of Invertebrate Zoology for Merseyside County Museums, for their help with the drawings.

PRINTED IN BELGIUM BY
proost
INTERNATIONAL BOOK PRODUCTION